AF885478

# Thou Shalt Not Think

## How Psychology Subverts Reason And Opens Pandora's Box of Faith

By Max Humana

Thou Shalt Not Think; How Psychology Subverts Reason And Opens Pandora's Box of Faith. 1st Edition

Published by Reality Publishing, LLC.
Copyright © 2018, Reality Publishing, LLC. Content is taken from the book *Thank god For Eve,* copyright © 2018, published in 2019.

All Rights Reserved.

For more information, contact Reality Publishing, LLC at realitypublishing2100@gmail.com.

Copyright infringement is against the law. If you believe the copy of this book you are reading infringes on the author's copyright, please notify the publisher at realitypublishing2100@gmail.com.

No part of this book may be reproduced, stored in a retrieval system or transmitted in any form or by any means, electronic, mechanical, photocopying, recording or otherwise, except for brief quotations, without express written permission of the publisher.

Cover design by Max Humana.

**ISBN: 978-1-7332857-2-8**

## Other Books By The Author

Thank god For Eve; Why Religion Exists And Why It Must Go

America as a Case Study In The Harmful Effects of Religion (Chapter Six from the book *Thank god For Eve* as a separate pamphlet)

An Opening Salvo On Coexistence; Plus Everybody's An Atheist And I Can Prove It (Chapter Eleven from the book *Thank god For Eve* as a separate pamphlet)

Religions Are Fruitful, They Multiply, And They Are Very Bad For You (Chapter Fifteen from the book *Thank god For Eve* as a separate pamphlet)

Why Is Any Moral Person Still A Catholic? (The Church of Evil, Abnormality and Hypocrisy) (Chapter Twenty-One from the book *Thank god For Eve* as a separate pamphlet)

America's Completely Secular Founding Documents (More Christian Attempts At Rewriting History) (Chapter Twenty-Seven from the book *Thank god For Eve* as a separate pamphlet)

Religious Freedom Laws, Also Known As "My god Says You're A Piece of Shit" (Chapter Twenty-Eight from the book *Thank god For Eve* as a separate pamphlet)

Four Degrees of Separation

# Dedication

This book is dedicated to all brave, courageous and happily godless freethinkers, humanists, secularists, scientists and supporters of science and reason in all times and in all places.

It is dedicated to every person who has ever advanced the causes of science and reason at the expense of dogma.

All these wonderful fellow humans deserve heartfelt credit and our collective thanks for their expansion of human knowledge. I give a special dedication to all the strong-willed women throughout history who have challenged the establishment and demanded that their equality be recognized.

From the first evidence-minded freethinking *Homo heidelbergensis* who raised a furry skeptical eyebrow at his fellow troopmate's obeisance to an imaginary entity, to our likeminded godless kin in the far distant future who will quite likely have to continue today's fight to keep religion and state safely separate, I salute you all.

Your courage in standing up for reason in the face of persecution and ignorance, your battles against mindless tradition and those who believe their god is on their side, and your never-ending pursuit of a better life for all creatures on this planet all demonstrate that imaginary gods only get in the way of doing what is right and what is good.

Religion's long war on reason and the reasonable has left long trails of tears and blood across our world. The image below depicts the kind of Inquisitional treatment which the deeply and sincerely faithful have so frequently and so eagerly inflicted upon fellow human beings who chose to think for themselves.

*Spanish Inquisition.*

The faithful have tortured and killed more people, and damaged more minds, bodies and lives, than we will ever be able to count. I would be inconceivably remiss if I did not sincerely dedicate this book to all of the men, women, children and the unborn who have lost their freedom and their lives at the willing hands of the devout.

# Contents

# Acknowledgments

All humans owe their very existence to their ancestors, and this existence is made flesh in the form of our bodies and brains. Our interactions with our parents, families and societies strongly shape our attitudes and perceptions.

Thus in the true spirit of genesis, I must first and foremost thank my parents for passing on to me the blend of genes which made me an empiricist and a realist. Fortunately, my inherited DNA orients me away from fear, superstition and its unjustified dogmas and toward courage, reason and an empirical reality.

I must also thank them, and primarily my mother, for providing an environment conducive to learning, thinking and advancement in the pursuit of education and intellectual self-development. The exclusion from our home of any requirements for dogmatic thinking and blind obedience to authority, combined with the freedom we had to explore ideas and information as a pure and unconstrained pursuit, set the conditions for many years of delightful reading, talking and thinking.

I also cannot thank my mother enough for the numerous instances of her selfless parenthood which I witnessed from my earliest years on this planet. My memories of these acts will remain with me until my very last conscious moment. Without ever calling attention to them, she set the example of what it means to actually do things for other people without any expectation of reciprocation, recognition or reward.

My mother was also instrumental in keeping me in the best schools she could find even when we could not afford them, and has been focused on education and critical thinking for as long as I can remember. I owe her, quite literally, everything I am, and I will always be grateful for it.

I must also thank my wife for her endless patience as I perused references, pored over transcripts, pounded away at the keyboard and spent time in dialogue with the faithful. And most of all, I want to thank her for her love and companionship.

# Thou Shalt Not Think

## How Psychology Subverts Reason And Opens Pandora's Box of Faith

*"But it's sad, people say, to be deceived. Not at all, it's far sadder* ***not*** *to be deceived. They're quite wrong if they think man's happiness depends on actual facts; it depends on his opinions."*
- Desiderius Erasmus

Most people think they are pretty smart. They think they are wise to the ways of the world and that their powers of perception are untainted by bias. They think they possess strong analytical skills which they sometimes admit may occasionally and in extreme cases be aided by a tiny bit of research. They think their current levels of knowledge, "common sense" and what passes for their critical thinking skills are sufficient to reach a clear and objective conclusion.

Most people are wrong.

What is worse, they aren't just slightly wrong, they are wrong to an astounding degree. Most people know essentially nothing about how the brain actually operates, from its biological framework to its psychological inputs and outputs to its internal processing. They also often know very little about the subjects on which they offer partially to wholly uninformed opinions. Despite this comprehensive lack of knowledge, they feel no embarrassment or hesitation about holding the opinions they do, regardless of how ignorant, uninformed or incorrect the opinions actually are.

And to paraphrase Haley Joel Osment's character from the movie The Sixth Sense[1], wrong people are all around us, they only see what they want to see, they only think what they want to think and they don't even know they're wrong. This has obvious consequences when people engage in a discussion on topics of any degree of importance. Even when the subject is something as inconsequential and ephemeral as the performance of the hometown sports team, people have become so animated in their defense of that team that they have actually assaulted and killed people over such trivialities.

When religion is criticized, and I am here speaking primarily about the monotheisms and particularly Christianity and Islam, the faithful still frequently respond the way they have since these religions were invented.

Only minutes elapse between criticism of these two faiths and the arrival of ad hominem attacks and death threats, with the most strident and unhinged threats being delivered by the most faithful and most sincere believers. In many places and times, more permanent and painful responses such as imprisonment, physical assaults and public murder of the criticizer are often the norm. Such unhinged lunacy is not some bygone aspect of the history of these religions. It happens all across our world today.

Try going to just about anywhere in the Middle East and publicly burning a copy of Islam's holy book, the Hadith or a caricature cartoon drawing of Mohammed and see what happens to you. The more deep and sincere the believer's faith, the more likely he will be to murder you. And in a world in which members of the two most widely held religions, Christianity and Islam, continue to threaten, attack, imprison and kill unbelievers, apostates and those of other faiths, we have a major problem.

Lest any overly smug Christians try to comfort themselves with delusions that Islam is the only religion with vast numbers of devotees who still act like this in the 21st century, all one need do is travel to any of the politically conservative states in America's south and Midwest and publicly criticize Christianity, any of its main characters or its teachings.

Most people are painfully aware of the typically murderous and destructive Islamic response to attacks on its faith because recent history is overflowing with examples of this. If you think Christians won't react similarly, try going to a public park or city street in Alabama, Texas or Mississippi and burning Jesus in effigy at the stake with a pile of bibles and consecrated hosts as kindling and see what happens to you. You will probably survive but you won't escape unscathed, and you're highly likely to stare down a gun barrel or fifty before you get away.

If burning mythological figures and bits of their reincarnated bodies in effigy isn't your thing, try loudly and publicly demanding actual evidence of the virgin birth, the resurrection, the Noachian flood, communications from the dead or the divine, or indeed of any miracles at all. Loudly proclaim that beliefs in such ideas are unsupported and rather silly childish dogma and that religion should be forever and permanently excluded from state activities.

In short order, believers from presidents and vice presidents to state and local politicians to televangelists to right wing media talking heads to clergy of all flavors and colors to mobs of Joe and Jane Six-packs (beer, not abs) will respond with horror at the 'grave offense' to their beliefs and will utter loud calls to rally the Christian Soldiers onward to defend the faith against all enemies, both freethinking and didactic.

As part of their desire to defend a long-dead semidivine Jew from the slings and arrows of outrageous logic and reason, it is a stone cold, rock solid and absolute certainty that some number of deeply and sincerely devout Christians, as previously noted, will deliver death threats to the "offender" and/or his family electronically, telephonically, via snail mail, package or in person.

This has always been the case with god's children, for the Christian lord their god hath repeatedly commandeth his followers in quite direct and unmistakable language to sticketh through with a sword or smasheth with stones anyone who questioneth their religion.

For those who are not familiar with that godly order to kill, it is in Deuteronomy 13 (KJV). I have added the bolding to the quote below:

> *"{13:6} If thy brother, the son of thy mother, or thy son, or thy daughter, or the wife of thy*

*bosom, or thy friend, which [is] as thine own soul, entice thee secretly, saying, Let us go and serve other gods, which thou hast not known, thou, nor thy fathers; {13:7} [Namely,] of the gods of the people which [are] round about you, nigh unto thee, or far off from thee, from the [one] end of the earth even unto the [other] end of the earth; {13:8} Thou shalt not consent unto him, nor hearken unto him; neither shall thine eye pity him, neither shalt thou spare, neither shalt thou conceal him: {13:9}* ***But thou shalt surely kill him****; thine hand shall be first upon him to put him to death, and afterwards the hand of all the people. {13:10} And thou shalt stone him with stones, that he die; because he hath sought to thrust thee away from the LORD thy God...*"

This of course comes only one biblical page after an earlier set of equally horrific divine orders which in the space of a few short sentences exposes the utter evil at the core of the monotheisms. The monotheistic god specifically tells his followers not to do what he specifically tells them to do in many other parts of the book.

He gives conflicting orders, and the vast majority tell the faithful to displace, harm or kill other people. Even when given a way out of the unparalleled cognitive dissonance the monotheistic books engender in people who are even remotely moral, the default is always to do evil to other humans.

Earlier in Deuteronomy, the order not to kill was buried in a large section of the book from which the Christian faithful derive their so-called ten commandments, the very mention of which fills their heads with insufferable arrogance and presumption. In Deuteronomy, the faithful are told their god will evict innocent people from their own land so the children of Israel can invade and occupy it. Again I have added bolding to some of the text for emphasis:

> "{11:22} *For if ye shall diligently keep all these commandments which I command you, to do them, to love the LORD your God, to walk in all his ways, and to cleave unto him; {11:23}* ***Then will the LORD drive out all these nations from before you, and ye shall possess greater nations and mightier than yourselves****. {11:24} Every place whereon the soles of your feet shall tread shall be yours: from the wilderness and Lebanon, from the river, the river Euphrates, even unto the uttermost sea shall your coast be*."

In the space of just a few pages, the faithful have been told to follow their god's own divine commandments, including one which specifically orders people not to kill (or do murder), then they are told to kill their actual brother, son, daughter, wife or best friend quite brutally and completely without warning.

As with so many shocking things in religion, this bears repeating. The faithful are told to immediately murder not invading armies, not strangers, but one's most beloved family members. You will note the order is of course addressed to men only, as husbands are not listed in those to be killed.

The faithful are told that their only recourse when being asked to worship other gods, the very first action they must take along with every other member of the tribe, is to immediately grab a sword or a stone and do their lord's evil bidding in order to ensure the miscreant wife, son or daughter is aggressively and thoroughly murdered. And their god will all the while be busily driving out innocent people from their villages so the children of Israel can have as much *lebensraum* as they want.

This sort of lunacy actually is quite consistent within the book, because this self-admittedly hyperjealous god has plenty to say on the relative importance of worshiping other gods versus killing one's own family. This is of course an activity at which this god is an old hand and at which he will repeat his prowess in spectacular fashion in the New Testament, often euphemistically described by apologists as "*Book II; The First One Was All A Big Mistake*".

Deuteronomy's commandments section, as opposed to the other set of commandments found in Exodus 20, or the third one in Exodus 34, says the following on these two activities:

On worshiping other gods, the guidance is:

> "*{5:7} Thou shalt have none other gods before me. {5:8} Thou shalt not make thee [any] graven image, [or] any likeness [of any thing] that [is] in heaven above, or that [is] in the earth beneath, or that [is] in the waters beneath the earth: {5:9} Thou shalt not bow down thyself unto them, nor serve them:* ***for I the LORD thy God [am] a jealous God, visiting the iniquity of the fathers upon the children unto the third and fourth [generation] of them that hate me,***

*{5:10} And shewing mercy unto thousands of them that love me and keep my commandments.*"

A full 98 words are committed to the orders and to describing just how galactically and generationally pissed off and vengeful the god will be if the no-other-gods order is violated and how badly he will punish offenders and their descendants.

On killing other human beings, all you get is "{5:17} Thou shalt not kill." That's it. Four words.

The apologist may try to say that the ratio of 98 to 4 is because the order not to kill is so simple. The realist will yet again note the continuous apologist denials of history and reason and say several things:

- Why all the frothing threats of just how badly the god will react to the faithful worshiping other gods, but there is none of this for those who would kill innocent humans who were allegedly made in the very image of the creator himself? It is clear this god actually has a perverted and psychopathic fetish for doing a lot of killing and for watching his own created people kill each other.

- Those who worship other gods are not only punished themselves, but their great-great grandchildren are punished too. These kids would not even in most cases be born until long after the original offender had died, and possibly as long as 70 years after the original "offense". This would make the ratio something like 392 to 4 (including each generation), and that killing is only 1% as important as whoring after other gods. This off-the-charts immorality is quite consistent with the biblical god's insane and murderous words and deeds.

- What we see is that this god really does think that murdering innocent people is about 1% as important a matter as his jealousy. This is borne out by all the commands to not only commit murder of strangers and beloved family members, but also to commit full-scale genocide.

- Why is there no explanation in any of the sets of commandments like "thou shalt not kill or do murder, except in these extremely specific and unfortunate circumstances, but then it's perfectly ok"? Even that bit of clarification would help matters tremendously.

- And why haven't the faithful been paralyzed over the conflicting guidance to kill and not to kill? 'To kill or not to kill, that is the question'. The faithful certainly have done absolutely massive amounts of killing in the name and service of their god, following his guidance despite and in the face of his own commandment which specifically orders them not to kill. Their god tells them to kill on many pages of their bible, then in a few snippets it tells them not to kill, and even then only in extremely few words with no explanation. One must assume that the sheer tonnage and volume of orders to kill have swamped the few isolated orders not to kill, much like the 392 to 4 ratio mentioned above but much worse in the direction of psychopathy.

- The truth is that the faithful often don't give a holy damn about any of those commandments or any of Jesus's teachings or place them in any special

> hierarchy or priority, except where and when it suits their personal preferences and goals. This is painfully obvious to the secular observer, but the faithful refuse to admit it. The disgusting and self-serving hypocrisy of nearly every self-professed Christian believer is absolutely appalling, and this is most glaring in America.

That last one applies to every bit of guidance the Christian bible offers. The commandments not to bear false witness, to steal or covet or commit adultery are routinely violated by adherents from allegedly infallible popes to priests to politicians to presidents to people on the street, whenever they feel like the desired ends justify the need to violate them. The same goes for the teachings of Jesus.

I will mention this many times, that most self-professed Christians in America today act like Jesus's teachings were things they should ***not*** do as opposed to very specific and direct things he said they definitely *should* do.

This is not a new problem. Mark Twain noted this hypocrisy a very long time ago when he said "*If Christ were here there is one thing he would not be - a Christian.*"

Since the beginning of their religion, Christians have used their holy book far, far more often as a justification for being inhuman to other people than as a reason to be good to them.

Earlier I said the god of the bible told his followers to commit rape. The evidence of that is in Numbers 31:1 and 31:2, when the allegedly loving but actually insanely jealous and angry god tells his henchman Moses to commit mass murder on the Midianites.

Moses, being a true believer and not so insufferably and arrogantly full of himself that he felt he should see his god's words as metaphor or requiring interpretation or being anything else besides the actual words, carried out the orders with great gusto. He and his men slew kings and commoners and did it so well there were no Midianite men left, since the book notes how they "*...slew all the males.*"

Along with killing every single Midianite man, the Israelites took plenty of captives and booty and left total destruction and scorched earth in their wake. The incident is described in some detail in the following quote from the King James version, again with bolding added for emphasis:

> "*{31:9}* ***And the children of Israel took [all] the women of Midian captives, and their little ones****, and took the spoil of all their cattle, and all their flocks, and all their goods. {31:10} And they burnt all their cities wherein they dwelt, and all their goodly castles, with fire. {31:11} And they took all the spoil, and all the prey, [both] of men and of beasts. {31:12} And they brought the captives, and the prey, and the spoil, unto Moses, and Eleazar the priest, and unto the congregation of the children of Israel, unto the camp at the plains of Moab, which [are] by Jordan [near] Jericho. {31:13} And Moses, and Eleazar the priest, and all the princes of the congregation, went forth to meet them without the camp. {31:14} And Moses was wroth with the officers of the host, [with] the captains over*

> *thousands, and captains over hundreds, which came from the battle. {31:15}* ***And Moses said unto them****, Have ye saved all the women alive? {31:16} Behold, these caused the children of Israel, through the counsel of Balaam, to commit trespass against the LORD in the matter of Peor, and there was a plague among the congregation of the LORD. {31:17}* ***Now therefore kill every male among the little ones, and kill every woman that hath known man by lying with him. {31:18} But all the women children, that have not known a man by lying with him, keep alive for yourselves***."

So the god of the bible and his *capo di tutti capi* on earth were hand-clappingly gleeful about killing thousands of Midianite men during the actual invasion. That was bad enough, but then after deliberate thought and in a display of full premeditated Hitlerian genocidal psychopathy, the lord's own soldiers followed their fuhrer's bidding, pulled their swords back out, which were still dripping wet with the blood of the Midianite men, and killed all the captured little boys, killed all of the non-virgin females including many mothers, and took the virgin women and girls as sex slaves and concubines.

It is one thing to read such unadulterated and absolute evil and be horrified and repulsed by it, particularly considering that these acts of pure evil are taken straight from the holy book of the god who is worshiped by over two billion humans today. It is quite another to pull back and think it through at the individual level and see it through the eyes of one of the victims.

Imagine you are one of the crying and terrified women and male and female children who are being held captive after seeing all of your fathers, grandfathers, husbands, older brothers and uncles ruthlessly murdered by an invading army. Your houses are on fire, your property has been stolen, even your animals are now captive with you, and you are huddling with the other survivors absolutely terrified out of your mind.

What can you as a mother, shivering with fear and having suffered the worst possible disaster, possibly say to your terrified children that can be of any comfort? All you can say is that even though all the fathers and grandfathers have been killed and the future holds no hope at all except to survive a life of slavery under horrible people in a strange place, at least you as their mother will be there for the children.

And then the crusading stormtroopers come back and start pulling mothers and women and boys out of the crying and screaming mass of prisoners and begin executing them in cold blood, on the orders of their prophet and his god. The murderers from Jehovah's army either begin a small-scale inquisition of the young women and girls to separate the virgin females from the "defiled", or more likely they took full advantage of the situation to sadistically "finger-test" all the girls to find the 'tokens of their virginity'. Those who were "felt" not to have these tokens were murdered alongside their mothers, sisters and little brothers.

The only survivors of these multiple rounds of deliberate murder are young virgin girls, from teenagers to terrified two-year-olds. These girls would have in short order seen the total destruction of their village, the murder of its men and their own beloved fathers, then had their mothers, older sisters and all their little brothers ripped out of their desperately clinging arms and brutally murdered in front of them, with many of the older girls only being murdered following at least digital rape by the murderers. This is as evil an act as is possible to inflict on an innocent person, and it all happened on the orders of the biblical god and his prophet.

Even if the apologist screams that this horrifically Hitleresque psychopathic story is some sort of metaphor or allegory, I dare any mentally and morally undamaged human being to come up with any positive meaning or implication behind the event whatsoever, or to deny its craven and total evil.

Note that what happened to the survivors was far, far worse than rape, although future rape of the young and innocent was clearly being ordered and supported here, from the god on down. Many survivors of acts of human brutality, from genocide to rape, have at least some sort of support structure to lean on while they are recovering from the trauma of being victimized, brutalized and raped. These young Midianite girls who were raped by the sons of Israel after the double rounds of Israeli genocide had no fathers, mothers, husbands, brothers or anyone else to go to for help.

They had no courts, no villages, no communities to protect them and help them recover. They were prisoners of war and sex slaves for evil men doing the will of their evil god and his evil "made man" on earth, and they had no one to turn to but their fellow genocide survivors and rape victims, most of whom were likely less than ten years old and themselves in desperate need of support.

And of course some significant number of the women who were murdered for the crime of having sex with a man were pregnant, in accordance with the god's command to be fruitful and multiply. Therefore the most holy Moses ordered his men to commit not only a second round of mass murder but also a mass abortion as a bonus. And the lord their god had absolutely nothing to say about it, therefore he was clearly extremely happy with it.

So yet again in this book of unadulterated evil we add one more mass abortion event to the list of mass abortions committed by or approved by the god which Christians worship so slavishly and sadomasochistically.

This is the god which billions of Christians worship. This is the guidance taken straight from that book of absolute evil they unbelievably call a "good book". This is why when they say they have the source of absolute morality we freethinkers and secularists recoil in disgust and horror. And they wonder why we nonbelievers want them to keep their evil religion and their damn book quite the hell out of our state and school business and extremely far from children.

And we can immediately dismiss any desperate apologetics by the faithful when they try to claim that any of these or any other words in their holy book weren't actually dictated by the god above, but came from some character in the book like Moses and are therefore not actually a dictum of the divine. They keep compounding their own mountains of evil and error by making such statements.

The fact is that if any scribe or prophet in the history of the planet ever got even a single word, phrase or quotation wrong anywhere in that entire collection of stories that were collated into the Christian bibles, the reason the errors are still in there today is because the god of that book was and is perfectly OK with all those errors being included.

If the Christian god has no problem with thousands of years of his one and only holy book being hopelessly garbled and full of errors and misquotations, and full of evil so demonic that it that would put Satan to shame, despite having all that time to make or "inspire" some badly needed corrections, then that is 100% his own fault. And, more to the point especially given his reputation for omniscience and omnipotence, it is therefore 100% his deliberate choice.

Either way, if one is a true believer, everything in the book should be taken as its god's direct or approved guidance and followed to the letter, or else the believer has risen to levels of arrogance only dreamed about by freethinkers, because the believer will then have to justify how and why he personally decided what his god got wrong and therefore should be ignored.

If I actually believed in a god as evil and vindictive as Jehovah I can't imagine having the stones to tell him he got anything wrong, because your punishment for misbehavior could be being turned into a pillar of salt, any or all of the tortures inflicted on Job, being divinely carpet-bombed, and/or of course an eternity of unmitigated full-body, nerve-destroying agony in a torture chamber of fire customized and personalized for each individual. It would be far better to be locked up for life or receive capital punishment while still alive on earth than to fail to follow the divine orders to rape and kill.

And again, one more time, and one hopes this is finally clear now, we can once and for all reject the pitiful attempt at diversion known as claiming the words somehow have to be "interpreted" or seen as metaphor. This is ridiculous and must be strongly called out for the utter foolishness it is. Christians say their bible contains the literal or inspired word of the Christian god or gods. It forms the foundation of the faith, and they take quite a lot of it at face value when they want to cherry-pick it to justify their actions.

We again return to the astronomical arrogance of the believer who says that he personally knows the mind of his god so unbelievably well that he can personally say which words from the holy book are to be taken as literal, which ones are metaphor or allegory and which ones can be safely ignored. Again, this god has had thousands of years to offer clarifying guidance on this subject and has done and said absolutely nothing.

Of course this is all nonsense. Nothing in the Christian or any other so-called holy book should be taken as anything but the ranting of ignorant and often aggressively hateful and disturbed men. The Christian bible in particular is hopelessly garbled, is full of errors and is nothing to take seriously except as ancient literature and a window into the demonic psychology of many of the authors and the barbarity of their time, and its use by millions of people to harm and murder innocent people.

All this ancient religious literature is just snuff fantasy written by people who knew nothing about the realities of life and our universe. There are indeed some truly useful things we can take from it, but all of them put together could fit on a handful of pages. The rest is nonsense, science fiction, historical fiction, mythology or evil.

And to dismiss yet another ridiculous attempt by empty-headed apologetics who in a last ditch effort will claim that the Old Testament was indeed quite bad but 'Jesus came along and made it all better', a quick examination of this claim will expose its complete fraud and illogic as well.

If you are a Catholic, you believe that Jesus and his father are glued together by some kind of supernatural ectoplasm called the Holy Ghost, and together these three somehow form both one collective and three separate entities at the same time. There has never in the history of the world been a coherent explanation of this bizarre dogma, and it is just one of many extremely bizarre dogmas dreamed up in that weird cult over the centuries.

Therefore in the Catholic sect the Jesus figure, being either a 33.33% part of and/or wholly the creator of the universe, is indeed the god of the Old Testament and has an equal share in every jot and tittle of its evil. Speaking of jots and tittles, it is Jesus himself who claims the entire Old Testament as his own, either by extension or by adoption, as quoted on the second page of the New Testament in Matthew (KJV), with these words; my bolding added:

> *"{5:17} Think not that I am come to destroy the law, or the prophets:* ***I am not come to destroy, but to fulfil****. {5:18} For verily I say unto you,* ***Till heaven and earth pass, one jot or one tittle shall in no wise pass from the law, till all be fulfilled****."*

So whether you are Catholic and a trinitarian or are not Catholic and are one of the other thousands of kinds of Christian, all your Christian god(s) are equally guilty of every evil in both the Old and New Testaments.

Thus we have simple and clear examples which show that logic, consistency and true morality are in no way a requirement or precondition for huge masses of humanity to become adherents of some religion or other. It is just as plain and clear that actual evidence is not required either. This can be easily proven by merely asking any believer to first cite the specific supernatural claims of his faith, then asking for actual empirical evidence for each of the claims, then also asking why he thinks the supernatural claims of every other religion that humans have ever followed are false.

Prepare to be galactically unimpressed at what will be offered to you in response, and even more dismayed at the weakness of human minds.

As of 2015[2], there are approximately 2.3 billion Christians and 1.6 billion Muslims on the planet. There are between 1.1[3] and 1.2[4] billion Catholics. That this is the case in the 21st century is both appalling and embarrassing for a species with the knowledge we have accumulated, and it indicates just how unbelievably widespread such completely broken human thinking is. This hyperabundance of adherence to the unbelievable, the unprovable and the immoral is a clear and present danger to the health, welfare and safety of every human alive and for generations to come.

This is the case because far too many of the faithful refuse to keep their beliefs to themselves, and busybodily and arrogantly go around in word, deed and in state and federal elective offices busily trying to inculcate their specific religion and its tenets into public law and ram it down everyone else's throats. We quickly see how important it is to bring every human mind to a common understanding of the processes, biases and flaws innate to human psychology. This must occur in conjunction with training on how to remove these biases from thought and action.

If you are a true believer, your biases which lead you to deeply and sincerely believe that your religion is not merely "little t" true but is actually the singular and unique One True Faith make you a contributor to that clear and present danger to the rest of us, as history has shown since the first religion was invented. You as an individual may not feel the need to persecute others, but your belief in unprovable supernatural propositions clears the way for anyone else to claim as true the supernatural propositions in his religion which require him to persecute and murder.

The bottom line is that no fully informed and productive discussion of beliefs, facts, opinions or any exchange of information and power-sharing agreement between humans is possible unless all participants are of sound mind and also deeply and broadly informed on human psychology. Since extremely few people meet this description, most human communication is fraught with misperception, miscommunication and the tragically foolish elevation of opinion and feeling over facts and reason.

It is for that reason that I will address these issues of the lack of understanding of basic human psychology and also of the problems of bias and broken thinking in this book. I do this to inform the reader of the existence of these logical breakdowns and the role they play in destroying communication and in blocking a common acceptance of reality. It is a tragedy that all human children are not being taught in detail how the human brain operates and how to identify psychological biases and traps, along with logical fallacies, both in themselves and others.

Thus of the seven and one half billion people alive today, only a microscopic fraction of one percent are actually aware of the facts of this serious problem. Therefore as a species we are collectively flying blind as we try to communicate, inform people of facts and how to handle and overcome emotions and uninformed opinions. If you think of our situation like that of an airliner in which all but one of the people on board are blind, and the seeing person is neither of the pilots, you have an idea of the scope of the problem and the gravity of the possible outcome.

Knowledge of these biases is only fairly widespread in a very small community of those trained or knowledgeable in psychology and psychological science, and since we are all human and subject to the unconscious demands of our unreason, even those most informed on biases and flaws in thinking are not immune to their effects.

No one is immune to the raw emotion that we feel when faced with moral dilemmas, challenges to our core beliefs or images or ideas that disgust or compel us. But the science behind all this is becoming increasingly well-documented by truly outstanding research and publications.

The problem rears its ugly head the highest when the faithful work so hard to forcibly inject their particular religion's tenets into other people's lives. Take a classic example. Right now, I want you to imagine two men having intercourse. Women vary greatly in their responses to such an imagining. Some get turned on by it, some are repulsed by it, and others occupy every step in between, as you would expect. For homosexual men this sort of activity is what they do in reality and what they think about, since that is a normal thing for their minds to do. It is thus a positive and pleasant thought for them unless they have been convinced by some religious person that these thoughts are shameful or evil.

For almost all heterosexual males, trying to imagine this situation is extremely difficult and unappealing, and is accompanied by some degree of discomfort up to complete and utter repulsion or disgust. The strength of the emotional response is largely not under the person's voluntary control, although nearly any response to any stimulus can be dulled through repeated exposure and desensitization, and a reasonable and mature person can control strong negative reactions.

The difference between the religious believer and the nonbeliever is that even if the visceral negative or disgust response is stronger in the nonreligious person, he will never seek to stop the two males from engaging in their activities through trying to invoke some religion to justify enacting some sort of law against it. He will not try to castrate, imprison or murder those who would engage in that activity.

The religious person, however, even if they don't try to murder or castrate such people, can almost universally be counted on to try to deny these people the same human rights they feel their god granted heterosexuals.

Their justification for this discrimination comes directly from their religion, but their motivation for it comes directly from their unconscious and unthinking bias and disgust, which have short-circuited their conscious critical thinking, enabled their hatred and tribalism and diminished or eliminated their morality. The words they use to justify their discrimination are religious blather, with no actual reasonable or justifiable empirically proven rationale.

Why are so many of the religious so quick to follow the commands given in their holy books to do bad things to gay people, but they are not nearly as fast to murder people? It is not because there is no clear biblical guidance on killing; you can kill anyone on earth including the unborn and you can easily find some kind of justification for it in the Christian bible.

It is because there are secular laws with severe penalties for killing people for preaching a different religion, but far fewer laws with far weaker penalties which protect the rights of gay people. It is also because the faithful scream at the top of their lungs that they deserve the right to discriminate against people they don't like because they say that preventing them from doing so infringes on their religious freedom.

Their religious freedom, they say, must extend to all people with whom they interact, which is just a level of assholery only a mindless bigot or deeply and sincerely religious believer can reach, and is absolutely not what the protections for religious expression were meant to cover. They were meant to protect people from being persecuted or worse for their religion, not to enable believers to persecute others or worse. This is again so typical of the faithful to use an original good intent and turn it around to use it to hurt others or defend their own thuggish hypocrisy.

The difference between the religious and nonreligious communities is that the nonreligious tend to leave people alone as long as they are not, in the simple but powerful words of Sam Harris, diminishing the well-being of conscious creatures. The religious, on the other hand, continually go many miles out of their way to try to impose their personal and religious dogmas and opinions on other people, blissfully unhindered by factual analysis, true human morality or any ethical restraint.

To continue the above example, why would a person want to stop two people who truly love each other, whether of different or the same sex, from getting married? The religious will proclaim that their holy book and the imaginary god which they think wrote or inspired it told them this was how they were to act. Taking a tour through the bloody and immoral sadistic disaster that is the Old Testament, one can indeed find that the allegedly loving and forgiving deity who allegedly made us in his own image hath indeed plenty to say or inspire about how we should handle homosexual behavior, along with a massive list of other activities. In the King James version of the Christian bible, Leviticus says:

> "*{20:13} If a man also lie with mankind, as he lieth with a woman, both of them have committed an abomination: they shall surely be put to death; their blood [shall be] upon them.*"

So the allegedly loving and all-forgiving Christian god feels that homosexuality is a capital offense. This is quite interesting, since humans are, again, said to be made in the god's own image, but for some reason he says male homosexuals just damn well need to be killed. One might wonder why lesbians are not mentioned here; this is quite likely because women were property and didn't matter a damn to the god or the book's authors.

While this biblical order to kill male homosexuals is quite clearly a psychopathically immoral statement, it is equally quite clearly the norm for this book. The faithful during biblical times were exactly the same as the faithful today in that they cherry-pick only certain bits out of their mythologies to justify treating other people badly based on behavior which they personally don't like.

And again, what the faithful never do, in a stunningly pervasive and never-ending display of peerless arrogance, hypocrisy and outright middle-fingering toward their deity, is follow the entire book. Apparently while the god's orders to kill gay people are to be followed, or if gay people can't be killed they should be subject to legalized faith-based discrimination, the rest of the orders in the book are conveniently ignored.

Thus the non-literalist faithful all give that stiffly erect middle finger to their god. Why a mere mortal true believer feels he can pick and choose which divine orders to follow and which ones to ignore has always been beyond comprehension. Such arrogance far exceeds any arrogance of which the faithful so often accuse the nonbelieving and scientific communities, by hundreds of orders of magnitude.

While the bible is full of enough immorality to choke a black hole, even just the rest of Leviticus has plenty to say on the subject of what the lord their god wants the faithful to do regarding many other behaviors. The following are just a few examples, taken from the very same part of Leviticus as the citation above. The allegedly loving creator of the universe wants his faithful to kill other people for:

- **Adultery.** *"{20:10} And the man that committeth adultery with [another] man's wife, [even he] that committeth adultery with his neighbour's wife, the adulterer and the adulteress shall surely be put to death."*

- **Incest, but only certain kinds.** *"{20:11} And the man that lieth with his father's wife hath uncovered his father's nakedness: both of them shall surely be put to death; their blood [shall be] upon them. {20:12} And if a man lie with his daughter in law, both of them shall surely be put to death: they have wrought confusion; their blood [shall be] upon them."*

  Note that this somehow did not apply to Lot, who screwed both of his own two biological daughters not long after his god, in a feat of divine sado-multitasking, turned Lot's wife into a pillar of salt while firebombing Sodom and Gomorrah. This bit of apparently quite acceptable god-sanctioned biblical family values comes from Genesis 19:30 to 19:36.

  Therefore Jehovah blesses the act of men screwing their biological daughters, but a grown adult having consensual sex with a member of the same sex is a capital offense. Such is the insane word of their lord. But even hetero sex is sometimes dangerous. Screwing one's daughters is perfectly OK, but sex with a woman and her mother means you need to be burned alive, as per the order below.

- **Sex with a woman and her mother.** *"{20:14} And if a man take a wife and her mother, it [is] wickedness: they shall be burnt with fire, both he and they; that there be no wickedness among you."*

As if the above weren't enough, the allegedly loving Christian god wants extremely draconian penalties for other behaviors. These behaviors are extremely heinous to the god who, yet again, is said to have made us in his image but is so disgusted by those bodies that they need to be kept covered up at all times. And menstrual blood is to be avoided like the plague, as it is apparently almost as toxic as the superacidic blood in the creatures from the *Alien* movie series. These other divine penalties result from:

- **A man seeing his sister naked.** *"{20:17} And if a man shall take his sister, his father's daughter, or his mother's daughter, and see her nakedness, and she see his nakedness; it [is] a wicked thing; and they shall be cut off in the sight of their people: he hath uncovered his sister's nakedness; he shall bear his iniquity."*

- **A man and woman having sex while she is menstruating.** *"{20:18} And if a man shall lie with a woman having her sickness, and shall uncover her nakedness; he hath discovered her* *fountain, and she hath uncovered the fountain of her blood: and both of them shall be cut off from among their people."*

"*Her sickness*", indeed. And the omniscient creator of the universe has spent thousands of years deliberately ***not*** clarifying that menstruation is not a sickness, along with a planetload of other information that would have been great to know, such as that slavery of any kind should be prohibited in all times and in all places, and that germs and not sin cause disease. So the god of the bible knew as much about science as a person who lived in the Iron and Bronze Ages…what an unremarkable and wholly damning surprise.

- **A man seeing his aunt naked.** *"{20:19} And thou shalt not uncover the nakedness of thy mother's sister, nor of thy father's sister: for he uncovereth his near kin: they shall bear their iniquity."*

- **Extended incest.** *"{20:20} And if a man shall lie with his uncle's wife, he hath uncovered his uncle's nakedness: they shall bear their sin; they shall die childless."*

- **More extended incest.** *"{20:21} And if a man shall take his brother's wife, it [is] an unclean thing: he hath uncovered his brother's nakedness; they shall be childless."*

Lot and his biological daughters did not incur any incest penalty, after he got both of them pregnant after screwing them shortly after his

> lord had killed his wife who was the mother of his daughters. This bit of biblical family values was just peachy with Jesus and his father. Having sex with your daughters is no big deal but having sex with your brother's wife is just going too far.

So why aren't all of the faithful scrambling and fighting with every inch of their lives to kill all adulterers? Or to impose draconian and permanent legal penalties and discrimination against all men who have sex with menstruating women, menstruating women who have sex and those who see members of their family naked? The faithful fight tooth and nail every single day to be able to discriminate against homosexuals. Why the galaxy-sized disconnect?

The answer is clear. It is simply that the faithful just have less of a personal disgust issue with heterosexuals committing adultery or screwing during certain days of the month than with homosexuals having sex, and also because they can get away with this particular bigotry but cannot get away with murder. At least not until the rise of a Christian State which would without a doubt certainly codify such truly insane penalties into law.

Particularly when adultery is corrosive to a relationship and often ends in divorces and broken homes, often with resulting negative effects on the divorced parties and their children, why aren't the so-called "family values" faithful coming out in droves and taking the case for criminalizing adultery and discriminating against the adulterous to the Supreme Court? There is quite simply no consistency with the religious and their actions, except their extreme inconsistency.

I have to caveat the above with a warning that while they do not operate widely or openly in the mainstream, at least not yet, there are indeed quite large and growing numbers of theofascists in America who are actually working toward implementing such Old Testament rules into secular law.

There are far too many people who would be very happy to stone adulterers to death and exile people who have sex during menstruation, and they would be quite happy to hypocritically cast those stones themselves. This strategy is being discussed and where possible the ground is being prepared for insertion of such laws into the local, state and federal levels.

Don't believe me? Go read Chapter Six of my book *Thank god For Eve*, which briefly covers the history of religion in America. Do some good solid research on your own, and when you are finished you will be appalled at what people are trying to do for their lord and savior. Hint – try "Dominionism" and "Christian Reconstructionism". And spend some time in fundamentalist churches in any Republican state in the country. You'll leave the service with quite a different view of religion, rising nausea and a burning desire to keep that wall between church and state absolutely and completely impenetrable, believe me.

The reality is that people feel very free to keep the beliefs they like, to discard what they don't want to believe, and to work their way through or simply ignore the rationalizations and cognitive dissonance necessary to arrive at some personally and individually acceptable faith and belief equilibrium. Every religious person deems some set of beliefs within his faith to be worthy of his attention and has another set of beliefs which he has heard or read about but which he feels quite fine with rejecting. Christians and Muslims all over the world go on endlessly about their so-called deeply and sincerely held beliefs, but offer no real evidence for the claims of the faiths. Believers really believe what they say they believe, but the inconsistencies in each faith and indeed in each believer would take decades to write down.

So how do the faithful actually arrive at the set of beliefs they choose to retain and follow? How do they decide which of the bits of their books they deign worthy of personal adherence and/or so galactically important that they have to try to make the rest of us follow them too? This is in fact the most important first area of inquiry we need to understand, because humans are very complex organisms. Logic is only one part of what we think and do, and it is often wholly absent.

Unfortunately, even when it is present and used, logic is applied improperly and/or is also often quite far down the list of considerations during belief formation and accretion. Logic is thus often ignored or rationalized away.

As Michael Shermer has said so accurately and so consistently, and as many others have suspected, known and said, our beliefs come first, full stop. Then, after we have the beliefs, we strive mightily to find ways to rationalize or justify them. We search around for what we perceive to be things that support our beliefs, and we ignore disconfirming facts and evidence. Or we don't even bother with trying to justify the beliefs, which is actually the case with many people, because for them the mere fact that they hold the beliefs is good enough for them, and they are sure it will also damn well need to be good enough for you and me as well, enforced by law if possible and handfuls of heavy stones if they can get away with it.

And there is apparently no belief so extreme, so ridiculous or so unsupported by evidence that large numbers of people will not still "deeply and sincerely" cherish and act upon it, even to the extent of willingly sacrificing themselves or killing others in the service of the belief. Many people believe things with every fiber of their being, despite the belief being wholly and demonstrably untrue. The difference between these people and the rest of us is that their faith is more powerful than their reason.

This type of person will not change his mind about a belief due to someone else's perfectly cogent, factual, informed and convincing argumentation, or even due to crystal clear evidence seen by his own eyes proving incontrovertibly that his belief is wrong. Reality does not matter to the deeply faithful and it never will. This is very hard for empiricists to understand, but it is as real as an Inquisitionist's torture chamber.

A person's level of investment in a belief drives his ability to ignore evidence against the belief. People who are heavily invested in and committed to a belief will increase their level of mental resistance as much as they need to in order to block the disproving evidence. That is one of the most important facts to remember about the ability of any kind of unfounded beliefs to take over weak minds. If a person has deeply and sincerely committed to believing that two plus two equals five, and this is justified due to some supernatural claim, no amount of facts or evidence will convince him that two plus two is four.

With regard to those who believe the claims of religion, their ability to ignore facts that disprove their religious beliefs thus strengthens in proportion to their level of personal and social investment in that religion and its claims. People cling to some beliefs particularly tenaciously due to their desperate need for comfort and certainty.

In some cases, this comfort can be essentially infinite, such as when a person believes in an eternal heavenly afterlife. Thus changing these beliefs would in practice require the believer to be faced with more than infinite evidence, which is of course impossible, so such beliefs are impervious to facts. There literally is no changing some people's minds.

The good news is that the body of empirical work on the faults that occur in human thinking unguided by rigorous external processes such as the scientific method is massive and growing by the day. One of the best descriptions which I have ever read on this subject from experts in the field of human cognition comes from Steven Sloman and Philip Fernbach, professors at Brown and the University of Colorado, respectively, in their work *The Knowledge Illusion: Why We Never Think Alone.*[5] On page 172 of this fine book they note that:

"As a rule, strong feelings about issues do not emerge from deep understanding."

The authors are being overly polite here. What this is really saying is that people who don't have a damn clue what they are talking about are the most likely to be ignorant stubborn horse's asses and to hold completely uninformed opinions on any subject whatsoever which wouldn't survive an elementary-school level critical analysis. This is more than an observation. It is a fact and it is well-known even to amateurs in the field of social interaction. This fact, and the damning and troublesome problems it causes, are not widely understood or accepted, although they are major factors in interpersonal and societal discord.

The rest of the page in *The Knowledge Illusion* from which the above citation comes is full of other outstanding tidbits, some which I have listed below as food for thought on this issue.

1. Regarding those strong opinions which do not emerge from deep understanding, in other words beliefs which are deeply and sincerely held by the ignorant and the stubborn, the authors quote two other personalities who commented on this subject. One comment is from the famous intellect and critical thinker Bertrand Russell, who said "The opinions that are held with passion are always those for which no good ground exists."

   I have to differ with Russell on this, though, in one regard. The tone of his comment was congruent with the ideas noted in the book, but the comment was not well-stated. Facts are facts and they are much different from opinions. When other people try to deny facts, or slant them to fit an ideology or opinion or hidden agenda, it is not just reasonable but a downright mandatory requirement to respond with passion and vigor to stop the suppression or denial of facts.

   When someone tries to deny the Holocaust, tries to deny people basic human and civil rights based on bigotry, tries to destroy our environment because he denies global warming or is beholden to a religious or political extremism, tries to deny that a specific political or religious action is factually and empirically wrong or immoral, or tries to justify anything at all because they "think" or "believe" a certain way about it, we have to act. Those who actually do value facts and reality must respond with passion and a loud voice to keep the queen of Reason firmly seated on her throne and to boot the court jester Belief outside into the cold rain.

2. They also quoted, of all people, Clint Eastwood, who said, in an interview[6] with Time in 2005, "Extremism is so easy. You've got your position, and that's it. It doesn't take much thought".

   So apparently even Dirty Harry's alter ego has a more logical and true assessment of this issue than the mindlessly faithful masses. I don't know much about Eastwood but he summed up the problem quite neatly. The rest of Eastwood's quote, which was not in the book, ended with "And when you go far enough to the right you meet the same idiots coming around from the left."

   This is also quite true. There are right wing nutjobs and left wing loonies, and if we all ignored a certain percentage of the wingnuts on the ends of the bell curve we would all be much better off. This does not exonerate all moderates, however. So-called religious moderates may not kill you in broad daylight if you try to convert them from Christianity, but their acceptance of unfounded claims leaves the door open to all religious beliefs and practices, without exception and whether they are “extreme” or not.

3. The authors did a very good job of laying out some of the ways in which most people lack critical thinking and are extremely ignorant on every subject it is possible to discuss. The information they cited was both surprising and depressing, and while I highly recommend readers peruse the book itself, I had to include one item as an exceptionally illustrative example.

   They referenced an Oklahoma State University survey which asked people if they thought that genetically modified foods should be required to carry a label saying so. About 80% of people said such labeling should be required. That wasn’t the problem. The face-palming follow-on was the problem. It said “But 80 percent also approved of a law stating that there should be mandatory labels on foods containing DNA.” Yes, DNA[7]; that is not a typo. Deoxyribonucleic acid. Apparently these people feel that the core of life itself is to be feared as much as Christians, Jews and Muslims have long feared menstrual blood.

   This was so ludicrous that I had to see the data for myself. A quick ride on the Google train led me to the Oklahoma State website with the Food Demand Survey[8] document and the anti-DNA sentiment in its full inglorious splendor. Page four of Volume 2[9] Issue 9, from January 16, 2015, shows that 39.09% of survey respondents support a ban on sugary sodas, 63.77% support calorie limits on school lunches, and yes, a full 80.44% of respondents support mandatory labels on foods containing the genetic code of life. This is even more idiotic than wanting a label on bottles of water to contain a dire warning that they contain hydrogen.

My initial reaction to seeing the survey was that of the comedy show *Futurama's* Professor Farnsworth when he was faced with similarly peerless levels of both ignorance and stupidity[10] and said "I don't want to live on this planet anymore.[11]"

Then I remembered this survey was done by Oklahoma State University, which is buried deep in the cellar of the religiously stultified antiscientific Inhofian[12] bible belt, and took some temporary solace in the hope that the survey was done only on Oklahomans, who as a collective body astoundingly chose specimens like Mary Fallin and Scott Pruitt to positions of state leadership.

Then I reached out to the survey team, who informed me in an email that "The survey is distributed to nationally (sic). 1,000 surveys are collected each month, with demographics weighted to represent those of the US population." That says it all. My next question is how can I sign up for those Mars colonies?

Such unbridled ignorance is staggering and dangerous. How is it possible that in the world's current greatest superpower, with enough resources to educate all of its citizens to a very high degree, that so many people can be so pathetically and damnably ignorant? How is it possible that in the 21st century such large numbers of people are so blissfully devoid of facts and equally blissfully devoid of the ability to use them? To say this is a severe problem is to massively understate the gravity of the situation.

While we absolutely need to function in our personal and professional lives and to govern our societies by the facts of reality, we have to understand that many people absolutely ***do not*** do this. We need to fully understand how people arrive at their beliefs and what can be done to make this process happen in a logical and reasonable manner, with logical and reasonable outcomes. We cannot afford to let unsubstantiated beliefs and uninformed opinions affect our fellow humans, animals and the environment, as they certainly will if they are not strongly challenged and turned away.

We never really could ignore these opinions and be intellectually honest, but with the power to wreak major catastrophe and to kill large groups of people in seconds now being wielded by single humans, our margin for error is now completely gone.

How the brain operates, and thus how the human personality is expressed, is at its core a set of involuntary processes and expressions of each person's basal psychological framework, which is itself derived from the brain's physical construction and environment.

The brain is very much like a computer constructed in a factory, and it operates according to a few major driving factors as described below. Damage, changes or inputs to any of these factors can and often do affect how and what people think and believe.

People often experience short-term and potentially lifelong changes in their thinking due to the effects of various inputs and conditions such as indoctrination, education, both positive and negative parental care and attention, training in critical thinking, traumatic experiences, traumatic thoughts, hallucinations, brain tumors, near death experiences, dreams or nightmares and damage due to substance abuse.

Left to its own devices, each developing brain will grow into a unique physical construct with a pattern of components and modules which operate according to that pattern. Everyone knows how differently people can think, and on some issues there can be as many different opinions as there are people.

The rest of this book is a bit lengthy and detailed but this is necessary because a short summary will not properly illuminate the scale and scope of the problems we face and the reasons for these problems. Sound bites and summaries will not convey what we need to understand in this issue.

Humans face extremely significant challenges with our onboard programming, and this problem is worsening as our species learns more information and as increasingly more people have to peacefully coexist. These problems in processing lead to logic failures and cognitive biases that damage and distort our thinking. Solving these problems starts with identifying these failures and biases and understanding how they impact our thinking.

A great deal of relatively recent study and observation shows just how many obstacles the brain puts in our path to empirically supported understanding of reality, and the vast amount of nonsense and clutter which the scientific method clears away.

I recommend readers go through the entire next section both deliberately and with a reflective mindset. As you read, take some time to reflect on both how the information enables the invention and appeal of uninformed opinion and ideas like religion, and how to reduce or eliminate these problems in your thinking and the thinking of those with whom you interact.

It is now time to explain how people arrive at their beliefs. To do this properly we must delve into the physical construction of the human brain, the conditions it experiences and how these factors impact our thinking. Beliefs arise from the following four sources and conditions:

1. **The structure and composition of the brain.**
2. **Physical changes or damage to the brain from physical causes.**
3. **Physical changes or damage to the brain from nonphysical causes.**
4. **How the brain responds to experiences and sensory input.**

## Belief Source 1: The Structure And Composition Of The Brain

Figure 1 below shows some of the brain's major physical parts and their general roles. The brain is our cognitive hardware and circuitry and is basically an organic computer. The organic hardware provides the physical structure on and in which the brain's functions and programming operate.

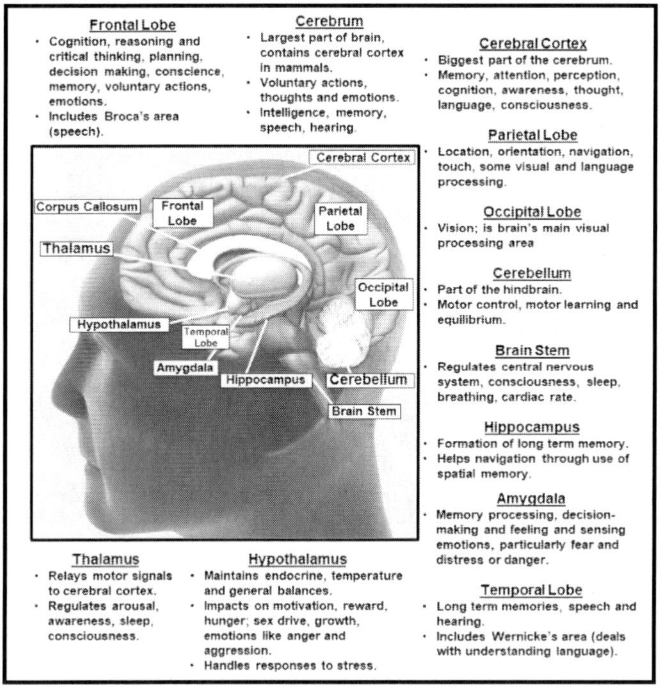


*Figure 1. Major parts of the brain. Main image courtesy of the National Institute on Aging, National Institutes of Health, U.S. Department of Health and Human Services; image adapted by the author.*

Blood and chemical-filled fluids[13] permeate[14] and surround[15] the brain and enable the hardware to operate. Those operations are the actions of neurotransmitters, firings of nerve cells and other activities which we have no idea are happening but which we perceive as reactions, emotions and thoughts.

If our brains and bodies were completely dehydrated we would lose about 80% of our mass and we would of course be quite dead. Thus it is the combination of the brain's physical structure, blood and those special fluids which enable the brain to conduct its conscious and unconscious functions.

Brain development[16] is an extremely complicated and very long process, which begins[17] on the 18th day of gestation and continues for the rest of the individual's life, though the major period of construction and programming is mostly completed between a person's 20th and 30th birthdays. This period is driven by genes and the physical environment in which the person develops. Genes and the environment will be familiar to non-DNA-phobic readers as the familiar twin pillars of nature and nurture. Development can be severely impacted or facilitated by negative or positive conditions in either pillar.

In the womb, the genes inherited from the parents drive the growth and development of the body and brain. If there is an error in the genes for brain development, the result will be miscarriage or birth of a baby with a malformed and/or malfunctioning brain. Note that this only refers to the physical development of brain structures, not the products or thoughts of that brain. We would certainly say from a moral perspective that a psychopath has a malfunctioning brain if we are describing the evil thoughts and likely behaviors of its owner.

But that person's brain may have developed in perfect environmental conditions and have perfectly followed its genetic code during assembly, and be completely psychopathic. We would call that brain well-formed in the sense that its physical development accurately followed its genetic blueprint with regard to the completeness and mechanical function of its parts, but we would also have to say that it was malformed and malfunctioning with regard to the morality of the thoughts it produces.

At birth, the human baby's brain has some software and an operating system built in. This is hard-wired into the involuntary or autonomic nervous system[18] which is itself composed of three sub-systems, called the sympathetic nervous system, the parasympathetic nervous system and the enteric or gastrointestinal nervous system.

Thus the combination of the newborn baby's brain and the chemical environment which saturates it are the equivalent of a computer's hardware and operating system, now plugged in and ready to accept input from the world outside the womb via the various senses, or peripherals. The baby's organic machine brain and nervous system include both reflexes and instincts.

According to Merriam-Webster, reflexes[19] are "an automatic and often inborn response to a stimulus that typically involves a nerve impulse passing inward from a receptor to the spinal cord and then passing outward from the spinal cord to an effector (such as a muscle or gland) without reaching the level of consciousness and often without passing to the brain".

Some examples[20] of reflexes are the act of jerking a hand back from a hot stove, or throwing up the hands in front of the face and turning the face away from an incoming object like a thrown ball. Babies have a host of reflexes including the startle reflex, the rooting reflex, the grasping reflex and the sucking reflex.

Instincts[21] are the "…largely inheritable and unalterable tendency of an organism to make a complex and specific response to environmental stimuli without involving reason." Instincts include the drive to eat, drink, have sex and intimate relationships, fit into social groups, seek sweet and/or fatty foods, defend yourself and your genetic kin and any allies, attack outsiders, gain higher social status, gain resources and seek enjoyment and entertainment. Instincts drive both good and bad behavior.

## *It's Time For A Revelation!*

It's time to discuss a situation which will be a revelation to many people. You probably already know that your reflexes and instincts are automatic parts of your subconscious, which is that "hidden[22]" brain that does a huge amount of work but of which you know little or nothing.

The same package of hardware and software which gives us reflexes and instincts also contains the patterns of our basic tendencies and predilections, thus our immediate and unthinking responses to stimuli. And also what we would call our personality. So far this is probably not too much of a revelation.

What will be a revelation to almost everyone is that nearly every human being's "hidden" brain processes have far, far more to do with how we think and what we agree or disagree with than we previously knew.

It is nearly a universal human characteristic that this hidden brain fundamentally determines our likes and dislikes, and thus how we think about and respond to subjects such as politics, religion, social issues like homosexuality, religion, climate change, science in general, people of other races or nations, who and what we choose to believe or not to believe, and every other issue we face.

This revelation means you are often nothing more than a slave to your "underbrain", and it is almost a guarantee that you don't have any idea that is the case. Odds are that even if you do know it, you don't work too hard against its wishes and are more than happy, even eager, to follow its lead. This is to be expected, because this is how you as an organic robot have been programmed to function. Acting in accordance with your programming is the natural path of least resistance.

Among a very rapidly growing body of truly excellent research into human cognition, two quite outstanding and relatively recent bodies of work deserve particular mention for their absolutely yeoman and long overdue work to illuminate this mental challenge we all have. The first is the work of the insightful psychologists Daniel Kahneman and Amos Tversky, and much of it is summarized in the wonderful book *Thinking, Fast and Slow.*[23]

The other and equally insightful and illuminating significant body of work on thought, emotion and opinion also sheds great light on what makes humans think and believe what they do. This other work, led by psychologist Jonathan Haidt and reflected in his book *The Righteous Mind, Why Good People Are Divided By Politics and Religion*[24], lays out much the same type of information but with other insights.

Haidt also notes how there is little conscious work done in forming our opinions and during the process of arriving at our beliefs. Beliefs that are comforting and "feel right" to us are the ones we keep, and those that don't "feel right" are rejected. And this is why so many people work so hard to defend their opinions, regardless of how little actual evidence exists to support them or how much contrary evidence exists.

There are two huge problems in human psychology. The first is that feelings matter more than anything to most people. The second is that facts only matter to those who choose to accept them. This strong tendency to avoid unwelcome truth and reality is why objective and peer-reviewed science is the only proven method to remove emotion and feelings when seeking real truth and knowledge. It is also why science must be the final arbiter of what is actually true and factual.

The two bodies of work by these insightful thinkers and their teams, underpinned not only by their own prodigious research but also by the quickly expanding list of congruent, supporting and relevant work done by others, neatly sum up the processes in the brain that make people react to the world around them in the ways that they do. They also describe the reasons why people first come up with and then cling so desperately to their opinions and beliefs, no matter how much logic or reality has to be twisted or denied in the process.

***"Thinking, Fast and Slow".*** Kahneman published this book fifteen years after his long-time research and intellectual collaborator Tversky died in 1996. The book describes how the brain's unconscious and conscious modes separately affect how we think and behave. It is as if every person has two separate personalities working independently of each other in the brain, with what we think of as ourselves in general control of one while being almost completely ignorant of and often a mindless slave to the other. Kahneman and Tversky described how the two modes of brain function separately act and react, and called these two different modes System 1 and System 2.[25]

**System 1** is the unconscious and so-called "fast" system described in the book's title. System 1 reacts extremely quickly and is reflexive and instinctual in nature. This means that as with reflexes and instincts, the system operates without deliberate or conscious cognition. The citation below describing the things System 1 does is taken from the first chapter of the book.

> "In rough order of complexity, here are some examples of the automatic activities that are attributed to System 1:
>
> - Detect that one object is more distant than another.
> - Orient to the source of a sudden sound.
> - Complete the phrase "bread and ..."
> - Make a "disgust face" when shown a horrible picture.
> - Detect hostility in a voice.
> - Answer to 2 + 2 = ?"

None of these activities require active brainpower and each one is essentially an automatic response, thus my reference to the action being reflexive and instinctive.

A critical point here is that emotion as an immediate and visceral response is included in this system. As everyone knows, emotional responses are absolutely not the products of deliberate analysis and careful cogitation, they are instantaneous and visceral, and they hit as hard as they are fast. You don't take time to think about how you will respond to a stimulus before responding; you reflexively or instinctively respond without thinking.

Imagine if you happen to be violently disgusted by images of internal organs, and were suddenly shown large and full-color mid-autopsy photos or even an actual partially dissected cadaver. Or that you are a devout chocophile and are presented with pictures and smells of Mom's best freshly baked chocolate chip cookies. Your immediate, unthinking and different emotional reactions to these very different stimuli, and the response you had to the earlier example of imagining homosexual sex, are the product of System 1. Your conscious "self" had very little to do with it.

**System 2** is the conscious, "slow" system. It involves deliberate cognition and thought and is not reflexive or instinctual. Examples of System 2 thinking include:

- Finding Waldo in one of those "Where's Waldo" cartoons.
- Doing a crossword puzzle.
- Keeping your eye on the player with the ball or the puck in a game like soccer, basketball, hockey or American football.
- Remembering the names of the major rivers in South America.
- Working out a complex math equation.
- Constructing a house floor plan to scale using pencil and graph paper.
- Answering a written question on a complex subject.

You can see how vastly different the systems are, and why this is the case. If you had to deliberately and consciously calculate and plan how to respond every time you experienced a threat like an object flying at your face at high speed, or the roar of a large carnivore behind you, humans would not be long for this world.

System 1 takes over in immediate response and reaction modes. System 2 can take over later, but even when it does, it is still largely underpinned and guided by System 1.

You are very often simply an unconscious slave to System 1, because you are an organic robot programmed to unconsciously respond to instincts and emotions. Your System 1 is always "on", and System 2 is like a computer in sleep mode. When you have to, you can wait for System 2 to wake up and see if it can help, but since System 1 immediately gives you what you think is an answer, i.e. it gives you a **feeling** and thus an **opinion** which normally turns into a **belief**, you most often don't ever bother to activate System 2.

This means you usually don't bother with consciously trying to validate information, and instead you operate on the uninformed and unthinking System 1 autopilot. Most people are lifelong slaves to System 1, and a lack of education often means this mental slavery is a lifelong problem. Those who suppress System 1 and engage System 2 before thinking and acting are unfortunately very rare birds, and are often dismissed by the System 1 slaves for "not having strong enough convictions". For a visual depiction of the difference between the two systems, take a look at Figure 2 below. It is the famous Muller-Lyer illusion[26], showing two parallel lines, one of which is capped by arrowheads pointing inward and the other with arrowheads pointing outward. Are both lines the same length?

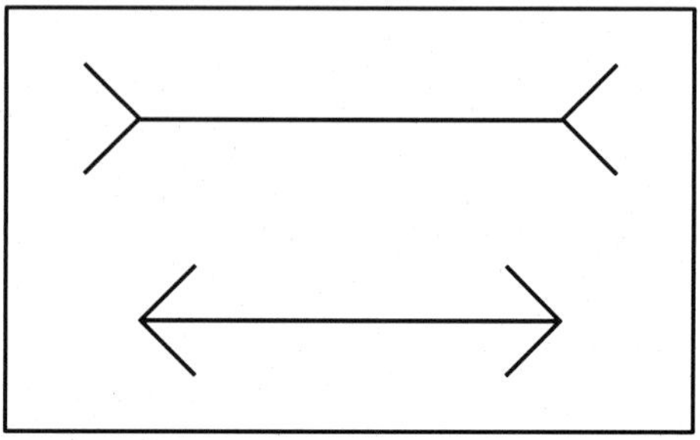

*Figure 2. The Muller-Lyer Illusion.*

Your fast System 1 will immediately say "of course not; the top one is obviously longer and my common sense proves it". Your slow System 2 will take the time to locate a ruler, measure the lines and say "despite how it looks, they are exactly the same length and the ruler proves it". The two systems operate separately and show how "common sense" is an extremely fallible concept and is often extremely wrong.

The Muller-Lyer illusion is merely one of countless examples of how common sense fails us. This illusion shows how observation and measurement should always supersede so-called "common sense" and opinion. In other words, empirical science and its method must always, in all times and in all places, permanently supersede faith when determining empirical truth and reality.

*Thinking, Fast and Slow* is over 400 pages of badly needed insight into human psychology, and I strongly recommend everyone who owns and uses a brain read the book and use it in every aspect of daily life, from interpersonal relationships to child rearing to voting, and to pass it on to friends and families.

I will summarize the book's major points without

going deeply into detail or the supporting science here, as readers really should invest the time to read the book for themselves. I have listed below some of the major takeaways from the book which are relevant to human beliefs in religion. The takeaway is listed first, followed by my linkage of its relevancy to religious belief.

1. **Priming.**[27] The book notes that "In the 1980s, psychologists discovered that exposure to a word causes immediate and measurable changes in the ease with which many related words can be evoked."

Religious relevance. At all stages in a person's life but especially during youth or times of stress, the words one hears in extremely frequently held religious ceremonies, meetings and prayers prime the brain to be cognizant of and more receptive to religiously-related terms and ideas. Note that there need be no actual facts of any kind involved, because in religion facts are absent. The words do the priming whether or not they are true. Religions work very hard to repeat their messages, and this endless and often daily repetition both primes and promotes reinforcement.

2. **Truth Illusions.**[28] When you make a statement that has one or more of the following characteristics - short, using simple words and easy to read, **in bold font**, **highly color-contrasted** from the background, and if it contains rhymes, it is perceived as more true. Again, it need not be true in any way, but it is still is perceived that way.

Religious relevance. Christianity is especially aware of this and is extremely vigorous in applying it, from nursery rhymes to poems to pithy sayings both spoken and on church billboards. Some examples of small-words rhyming Christian propaganda-messaging are:

- "Now I lay me down to sleep, I pray the lord my soul to keep."
- "In god we trust."
- "50 shades of grace."
- "Pie in the sky when you die."
- "Jesus died for your sins."
- "What would Jesus do?"
- "We aren't just molecules in motion."
- "From goo to the zoo to you isn't true."

These statements are cognitively easy, they flow and have a message. They thus appeal strongly to System 1 which, as we have already established, quite often overwhelms the logical System 2.

3. **Math Challenges**. Most people fail, and sometimes fail quite badly, when they try to do probabilities or statistics calculations in their heads.[29] This is because almost no one has a decent Probabilities or Statistics module in either our System 1 or our System 2, and therefore System 1 has to make an instant guess. This usually resembles a toddler trying to quickly answer a question on quantum mechanics. Probability and statistics are inventions of advanced mathematics, and human evolution has not had the thousands to millions of years it would take to naturally evolve well-functioning innate modules for them. Thus when we try to do probability and statistics without the aid of math and the appropriate training, most people get them very wrong because they default to their so-called "common sense" as a guide, and this

particular guide is not only often blind but deaf and dumb as well.

Religious relevance. When the faithful say that it is "too improbable" that some event could have happened without divine intervention, such as the evolution of the human eye, the origin of life or the spread of their particular religion, their numerophobia is expressed in the fear and stupefaction they feel toward very large and very small numbers. The faithful fall into a very deep psychological trap due to this inborn mental blind spot.

They are also extremely likely to have little to no actual knowledge and experience in the relevant scientific fields and histories, and less training in probability or statistics. Most things in life are not at all explicable by innate or inborn "common sense", and very improbable things happen all the time. A proper education and properly done math and science have a way of making even the fuzziest questions and answers spring into sharp focus, even if the final picture isn't the one you expected or wanted.

4. **More Math Challenges.** Their observation of how people assess the believability of stories showed that what people feel is true based on so-called common sense is often completely wrong and is in the exact opposite direction from reality. They thus "…introduced the idea of a conjunctive fallacy, which people commit when they judge a conjunction of two events…to be more probable than one of the events…in a direct comparison."

For example, this means that people think that a story about one person which lists multiple characteristics of that person is somehow more probable and believable than a story about the same person which only lists one characteristic. The truth is that every additional listed characteristic makes the story much *less* statistically probable, thus System 1 is 180 degrees out of line from reality and again common sense is dead wrong.[30]

Religious relevance. In Christianity, this means that people think that Jesus having only the characteristic of being born of a virgin with no other fanciful characteristics is less believable than after all the other fanciful characteristics have been added to the story, such as being the son of the creator of the universe, able to heal the sick and walk on water, able to rise from the dead, able to cast out demons and so on.

Each added fanciful characteristic makes the Jesus figure mathematically much *less* probable and believable, but the mind, particularly a gullible and less educated mind, erroneously and perversely thinks the opposite. Thus in a 100% contradiction of reality and logic, the crazier the story gets, the more believable it gets in many people's minds.

5. **Knowledge Gaps**. While it is impossible to become an expert in everything, many people are highly trained experts and very seasoned professionals in their fields. These people can reach a level at which they can have 'skilled intuitions', which are intuitions that are rather more probably correct than the intuitions of someone without the same level of expertise and experience.

This level is said to be achieved after a skill is reached through "an environment that is sufficiently regular to be predictable" and "an opportunity to learn these regularities through prolonged practice.[31]"

Religious relevance. This goes back to strong opinions coming from ignorance. Regarding religion, if a person were to become a true expert in all the fields of history, philology, biology, chemistry, physics, archaeology, psychology, anthropology, sociology, meteorology, astronomy, geology, economics, military science, genetics, neuroscience and evolutionary biology, then a person could actually have skilled intuitions about the claims of religion. Since the empirical facts of all these fields show that religions are wholly manmade, truly skilled intuitions will deliver that conclusion. The intuitions should nevertheless of course be followed by a rigorous scientific and mathematical analysis of the facts.

Note that believers and theologians are *very* different from actual empirical experts. Believers and theologians do not have skilled intuitions because they rarely have more than a passing awareness of the scope of any of these fields of study, much less deep understanding of any or all of them, including their own faith. And theology, being merely a study of a sliver of ancient literature, mythology and historical fiction, counts as an applicable field of study with regard to the real world only to the extent that empiricism can illuminate. That said, believers and theologians have fears and feelings in great abundance, and those are what drive their faith.

Seasoned experts in the study of ancient literature and mythology would have the potential ability for skilled intuitions in these fields, but it is impossible to be an expert or skilled in something that cannot be measured, such as the supernatural claims religions make.

6. **The Power of Fear.** They noted that people do not respond in the same way to potential gains or losses, and noted "We concluded from many such observations that "losses loom larger than gains" and that people are *loss averse*.[32]"

Religious relevance. This is highly reminiscent of the logic-twisting self-justification seen in Pascal's Wager. A believer who feels that his religion actually clears his path to heaven and eternal life is extremely likely to go on believing in that fantasy, because he feels it is a literally priceless reward which he will eventually receive if he can just keep believing. He thus handles his faith as an already-owned asset which he will go to great lengths to hold onto, regardless of the astoundingly broad denials of reality and rationality this will require.

Therefore asking him to logically assess his faith and take a chance on losing it is, to him, proposing a massive loss of eternal and infinite proportions. Therefore to avoid such a perceived loss, the believer's mind will take whatever twists and turns it needs to in order to remain in the faith. This is of course a generality, as a few people do have the strength and courage to leave their faith once they've invested in it, but they are rare.

While these are great specific examples of broken thinking, this book is well-populated with others, with each one a reminder that people are born with certain response patterns and are extremely susceptible to their own personal experiences, thus are psychologically "primed" to think a certain way. As a result people are very challenged when trying to "think outside the box", and this is why people who actually can think outside the box are so rare. Everyone's box is the cumulative action of both System 1 and System 2, continually updated by the experiences which every human collects over a lifetime.

If you live in a newly-discovered hunter-gatherer clan and you have never seen a flashlight, upon seeing one for the first time you will try to understand the workings of that wondrously magical object through the lens of what little you know. This is likely to be fire, the sun, the moon, the stars or whichever gods you worship. Your opinion will be that the light comes from a little fire or a piece of a celestial object harvested from the skies, or it is a magic gift from your god, probably a sun god. You will have no conception of batteries, wires or electricity. To truly understand how a flashlight works, you will first have to be taught about all those things, and also that there is no supernatural entity or action required to produce the flashlight or the light it creates.

In other words, the hunter-gather will say that of course common sense proves that his godly answer is the correct one, and only an idiot would say otherwise. This is why opinions and so-called "common sense" by themselves are often not just worthless but very counterproductive, and this helps to demonstrate why science is the only path to true understanding.

This sort of hyperconstrained thinking is why we continue to hear the ignorant arguments from the faithful about there having to be a god who created the universe or magicked life into existence, and why these special pleadings won't go away anytime soon. The human race, while having climbed both out of the primeval sea and a fair bit up the Mountains of Modernity and Knowledge, still has ideas and models for but no actual evidence from before the Big Bang.

We have no heuristic, as Kahneman and Tversky would say, and no experience of a time or place before the Big Bang, so there is quite literally an infinite number of possible ways to describe that time. We simply don't have the concepts and data yet to properly describe it, although we can use our current heuristics as a baseline. We are as ignorant of the pre-Big Bang period as those hunter-gatherers are about quarks.

What we do have as a civilization, however, is a continuous and unbroken series of now-dead religious explanations for a whole host of questions that have been wholly replaced by scientific explanations, with not a single explanation ever going in the opposite direction, and there is absolutely no reason to believe this will ever change.

Regarding that primeval explosion, it could be that "our" Big Bang is only the most recent of a quintillion such events in a series, with every single preceding universe being composed of the same matter and energy of this one, with no or only miniscule variations in its organization and distribution, and that this one happens to be different in that it is apparently going to expand forever[33] due to some pattern of distribution of dark energy instead of becoming sucked back into a Big

Crunch.

This one-off, one-in-a-quintillion variation could be due to a mutation in the universe-replication process, to bring a natural selection heuristic into the discussion. It could be, in all fairness, that some god or creature did indeed sneeze the Big Bang out of its ineffable breathing spiracles, although there is zero evidence of any such entity or of any such super sneeze.

The infinity of possible pre-Big Bang conditions provides the faithful with an infinite number of gaps in which theists can hide the specific god(s) of their personal choice, and this is exactly what they do.

To wrap up the work by Kahneman and Tversky, their findings are extremely informative and I again highly recommend that everyone who owns a brain read that fine book.

I also recommend all brain owners read the outstanding work by Jonathan Haidt and his fellow researchers. Haidt is a social psychologist and Professor of Ethical Leadership at the Stern School of Business at New York University. His work intersects in many places with the Kahneman and Tversky work and illuminates the same basic conclusions. He and his team have also gleaned insights that are critically important in understanding why people think and believe what they do.

This body of work on human reflexive responses to situations or scenarios gives great insight into the practical applications of understanding and dealing with human tendencies and frailties. Haidt and the team conclude, and I completely concur, that mental response and reaction patterns are set into every newborn baby's brain, and that these patterns drive who people choose to interact with or not, and how to think and act, or not. This process occurs throughout a person's life and as he feels appropriate within his local social and cultural groups.

These initial response patterns, unless changed by major physical or psychological impacts, drive the morality or lack thereof, politics, gullibility, aggressiveness, studiousness and religiosity of the growing person, along with the rest of the person's preferences, likes, dislikes and other tendencies.

While the work is full of outstanding and illuminating facts, points and sound rationale, there are two takeaways I want to specifically note from Haidt's work. I pulled both of them from his already-mentioned book *The Righteous Mind: Why Good People are Divided by Politics and Religion.*

The first takeaway is that he has done great work and conducted much external communication on the nearly universal human inability to come up with a reasonable and empirical justification for the positions people take on moral questions and situations. The relevance to religion will be made clear.

The second takeaway is that it is essentially a human universal that conservatives, liberals, the religious and the nonreligious all tend to strongly respond to what Haidt calls moral foundations, captured in "Moral Foundation Theory". This indicates which values are most important to people and how these values drive their unconscious and often inexplicable opinions on morality.

I don't necessarily mean their opinions are always wrong; the point is that people just often have no real idea why they feel the way they do about specific words and actions and often cannot explain the justification for their opinions at all.

The method they took to illuminate this human inability to justify personal moral positions was simply asking people if they thought the actions in some interesting and nuanced scenarios were moral, and to state why or why not. The questions were quite inventive and I have listed some of them here.

1. "A family's dog was killed by a car in front of their house. They had heard that dog meat was delicious, so they cut up the dog's body and cooked and ate it for dinner. Nobody saw them do this.[34]"

2. "A man goes to the supermarket once a week and buys a chicken. But before cooking the chicken, he has sexual intercourse with it. Then he cooks and eats it.[35]"

3. "A woman is cleaning out her closet, and she finds her old American flag. She doesn't want the flag anymore, so she cuts it up into pieces and uses the rags to clean her bathroom.[36]"

4. "Julie and Mark, who are sister and brother, are traveling together in France. They are both on summer vacation from college. One night they are staying alone in a cabin near the beach. They decide that it would be interesting and fun if they tried making love. At the very least it would be a new experience for each of them.

Julie is already taking birth control pills, but Mark uses a condom, too, just to be safe. They both enjoy it, but they decide not to do it again. They keep that night as a special secret between them, which makes them feel even closer to each other.[37]"

5. "Jennifer works in a hospital's pathology lab. She's a vegetarian for moral reasons – she thinks it's wrong to kill animals. But one night she has to incinerate a fresh human cadaver, and she thinks it's a waste to throw away perfectly edible flesh. So she cuts off a piece of flesh and takes it home. Then she cooks and eats it.[38]"

As you can imagine due to your own innate unconscious responses / moral foundations, most people felt that these acts were immoral or disgusting to some degree. For example, Haidt notes that only 13% thought Jennifer's cannibalism of the dead body was alright. (Feel free to insert your Catholic consecrated host joke of choice here). But people had a devil of a time explaining specifically *why* they thought it was immoral. Haidt used the word "flailing" when describing the attempts subjects made to justify their opinions and this seems to be the perfect word to describe the situation.

These interactions revealed the Kahneman-Tversky System 1 in operation, with System 2 trying to catch up but basically flapping in the wind and unable to say anything useful.

System 1 enabled test subjects to have an immediate opinion, but even after deliberately engaging their System 2 to answer the question of *why* they felt how they did, they couldn't make it come up with a reason for their opinion. In Haidt's clear and extremely relevant words, "Reasoning was merely the servant of the passions, and when the servant failed to find any good arguments, the master did not change his mind."

The relevance to religion, as well as to politics, is crystal clear and it has significant implications. Despite the total lack of justification for their opinions, Haidt's test subjects were nevertheless very often quite sure of those opinions. Thus the religious and religiously influenced opinions and feelings people hold, which form the basis for their unjustified religious beliefs, are embedded so deeply in the System 1 parts of the brain that even a highly engaged System 2 is often incapable of doing anything but short-circuiting and doubling down on what System 1 was feeling.

This mental challenge is why when people are shown disconfirming evidence for their opinions or beliefs it is often like waving a flag in front of a bull. The bull gets mad and tries like hell to gore you and get you out of his face, and the believer not only won't face facts or change his opinion, he fights even harder to defend his unjustified opinion, his unproven beliefs and his religion.

When one learns of this tragic facet of human thinking, one suddenly sees the endless, nonsensical and diabolically cruel biblical proscriptions in a new and different light. The biblical authors felt personally disgusted by certain actions, so they felt they were immoral or sinful, thus they felt such actions should incur severe penalties. Therefore these actions were listed out in painful detail and punished whenever they were discovered.

One must layer onto this already hyperignorant sadism the fact that these people actually believed that some other person's sins would result in godly punishment or vengeance not just on the miscreant but on the entire tribe. Thus the abject ignorance and immoral barbaric sadism that absolutely saturates the monotheistic holy books is at least partially explained.

The other major point I wanted to mention from Haidt's work is the implications Moral Foundation Theory has for belief and religion. Much as Newton's prism split light into its component colors, Haidt and his team's splitting of the unconscious mind's operations into the subelements and expressions of morality expands the general idea of the System 1 we discussed above into components of what Haidt calls "intuitive ethics".

Their work does a very good job of explaining psychological tendencies and what people think are moral and immoral, and a few of the biological reasons for these tendencies. But the main point is the theory itself, which he explains in the book and which you can also find at their website at http://moralfoundations.org/.

Haidt makes the point that these are not characteristics of just one culture or region or country; they are characteristics of the species *Homo sapiens*, as discovered through extensive research. Based on how they behave, it is a near certainty that to some degree the theory and its moral foundations are also relevant to much mammalian life on earth and particularly to the primates.

The theory notes that "intuitive ethics" is at the core of human psychological programming and processing, and that at least five “foundations” underlie the theory. The following description of the five core foundations and one other probable foundation is taken directly from the website cited above.

> “**1) Care/harm**: This foundation is related to our long evolution as mammals with attachment systems and an ability to feel (and dislike) the pain of others. It underlies virtues of kindness, gentleness, and nurturance.
>
> **2) Fairness/cheating**: This foundation is related to the evolutionary process of reciprocal altruism. It generates ideas of justice, rights, and autonomy. [Note: In our original conception, Fairness included concerns about equality, which are more strongly endorsed by political liberals. However, as we reformulated the theory in 2011 based on new data, we emphasize proportionality, which is endorsed by everyone, but is more strongly endorsed by conservatives]
>
> **3) Loyalty/betrayal:** This foundation is related to our long history as tribal creatures able to form shifting coalitions. It underlies virtues of patriotism and self-sacrifice for the group. It is active anytime people feel that it's "one for all, and all for one."
>
> **4) Authority/subversion:** This foundation was shaped by our long primate history of hierarchical social interactions. It underlies virtues of leadership and followership, including deference to legitimate authority and respect for traditions.

**5) Sanctity/degradation:** This foundation was shaped by the psychology of disgust and contamination. It underlies religious notions of striving to live in an elevated, less carnal, more noble way. It underlies the widespread idea that the body is a temple which can be desecrated by immoral activities and contaminants (an idea not unique to religious traditions).

We think there are several other very good candidates for "foundationhood," especially:
**6) Liberty/oppression**: This foundation is about the feelings of reactance and resentment people feel toward those who dominate them and restrict their liberty. Its intuitions are often in tension with those of the authority foundation. The hatred of bullies and dominators motivates people to come together, in solidarity, to oppose or take down the oppressor."

It is again clear that this work has strong and direct implications for religious belief. Once a religion is created, and people start believing in its tenets and a related social group starts to form, several of these foundations start to kick in, particularly loyalty/betrayal, authority/subversion and sanctity/degradation. Religion has both supernatural and social components. Thus some of these foundations actually have double impact in that they affect both components.

In point of fact, there is actually a third level of impact, because religions either explicitly or implicitly see themselves as the sole owners of a universal moral code, so a believer's ties are to the god or gods, to the group and also to the assumed moral superiority of the religion. Thus true believers are embedded in and wedded to their faith quite viscerally and emotionally with several independent but related ties.

The loyalty/betrayal foundation links to religion in each of those three ways. The practicing true believer has to be loyal to the god, loyal to the clergy and loyal to the moral code. Betraying any of these is considered a grave "sin", if you will, and the faithful feel that it must be punished. Punishment can be inflicted in this world through expulsion, excommunication and/or execution, and in all likelihood still more punishment will be required in the afterlife, applied directly by the god of the faith, one of his minions or some devil or demon.

The authority/subversion foundation's inherent tie to religion also uses all three linkages. Believers must respect the authority of the god, of the human leaders of the social group/religion, and the assumed superior authority of the religion's moral code. If you like, you could add a fourth linkage, which would be the traditions Haidt mentions. Thinking you are still a practicing believer but refusing to partake in rituals and traditions will almost certainly mark you as a subversionist and a strong candidate for at least some degree of ostracism if not full expulsion.

In the extract cited above, Haidt specifically named how the sanctity/degradation foundation links to religion. This particular foundation is *extremely* relevant to religion. The faithful doggedly pursue a utopian Never-Never Land in which all sin and misbehavior are banished, and in which the human body and soul are cleansed of all things considered disgusting and unclean by the brethren and their gods.

The problem is that this also includes even the thoughts of doing things that are sinful, and depending on their degree of sadism and masochism humans will call almost any thought or deed sinful, especially if it involves sex.

Therefore some particularly abnormal and cruel people will say that even the most completely normal and/or innocuous human desires and actions are disgusting and require shame, guilt and extreme punishment.

All the foundations set the stage for inclusiveness or xenophobia, for acceptance or hatred, and for both moral and immoral acts. If a person responds too strongly to any of these foundations, people who are in his group will be accepted and those not of his group will be treated with fear, disgust, distrust and contempt. This sort of response can be merely a snobbish internal and unspoken disdain for other people, but it is too often not so constrained or hidden.

This kind of thinking and acting was raised to cataclysmic proportions in the Final Solution, in which Hitler and his largely Christian followers saw Jews as a disgusting and not fully human "race" of Christ-killing vermin who should all be murdered.

The hatred and bigotry people feel and act on towards people have many names, including racism, sexism, misogyny and homophobia. People are feared, hated and often mistreated for being of a different ethnicity, color, gender, religion or lack of religion, nation of origin, social class, high school or college, favorite sports team, and indeed any other characteristic that differs from one's own group.

The problem is that almost every human alive feels some degree of xenophobia or distrust in some form or fashion. Some people are hypertribal, while others are merely uncomfortable with newness and change. Some humans are the opposite, and are hyperinclusive with no, very few and/or very specific hypertribal tendencies.

The conservative mind, which is often also a religious and relatively dogmatic and inflexible mind, is more xenophobic than average. The liberal mind, which is often less religious and more flexible and nondogmatic, is less xenophobic than average.

We are all generally programmed to have tendencies like this due to many millions of years of evolution and through some artificial selection. We are specifically programmed to have our own personal tendencies based on individual differences in brain structure, social programming and other influences as described earlier in this section.

Liberals may think that conservatism is actually a pathology. Conservatives may think that liberalism is actually a pathology, but in the opposite direction. The truth lies at the extremes and in the context.

As in all observations humans make, the less populous but much louder extremes always stand out more than the quieter and more moderate masses in the middle. Just as the squeaky wheel gets the grease, and disasters, terrorists and disturbing or abnormal outliers get the "if it bleeds, it leads" headlines and attention, the raving lunatic right wing nut jobs and far left wing uberpeaceniks get the attention.

Hyperconservatism, when applied to human societies and human interactions, is indeed a pathology. Before liberals get too excited, they should note that hyperliberalism is also a pathology, when applied to other aspects of human interaction such as individual and group safety and security.

Conservatism, in the classic sense of tending to preserve the current order and having a reasonable caution about new things, is a necessary part of a reasonable and balanced approach to human thinking and action. Note that this does not in any way refer to the American Republican party of 2016 and the last few decades, as it has quite deliberately destroyed any sort of reasonable or rational "little c" conservatism.

Merely being "little c" conservative is not pathological. Republican politics in America in 2016 is highly pathological.

Liberalism, in the "little l" general sense and not referring to the American Democratic party at all, but in the general sense of openness and acceptance, is also a necessary part of properly balanced human society and thinking. Minor differences are necessary to cleave a reasoned solution from a tangled web of reasoned "little c" conservative and "little l" liberal perspectives.

Hyperconservatism is a pathology.

Hyperconservatism is a hyperexpression of the fear of outsiders and disgust at contamination, which expresses itself in xenophobia and sadism in a horrible and often murderous blend. Fear of others served an important role in our evolution because such behavior kept invaders away and kept the tribe safe from all intrusions, be they dangerous diseases or dangerous ideas. Keeping unclean things away prevented disease. Keeping unclean people away prevented crime, misbehavior and challenges to authority. Keeping inconvenient realities away and having far too conservative a world view has led to mass persecutions and killings throughout human history.

Mistreating, demonizing, persecuting and even killing people merely because they are different is evil and immoral, but history proves that it is an inevitable result of hyperconservatism. Religion is usually ubiquitous in both conservatism and hyperconservatism, and it serves as a blanket justification for the entire spectrum of evil acts. The holy books of the Mesopotamian monotheisms, and the religions they spawned, are perfect examples of hyperconservativism, hypertribalism and hyperxenophobia.

Hyperliberalism is a pathology.

Hyperliberals love everyone and everything and don't quite get that there are psychopathic individuals who are born that way, who are not treatable or even easily identifiable with current medical technology, and who will do their damnedest to hurt and/or kill people. Hyperliberalism opens too many doors to infection by disease and to being conquered by aggressors. Liberalism and acceptance of outsiders, though, served a role in our evolution because new people and new ideas prevented things like inbreeding and damaging mutations from having too-small gene pools.

Liberalism and acceptance enabled the invention and incorporation of new technologies and better or more efficient ways of living. Without new ideas and some degree of acceptance of new people and the entire concept of progress which is inherent in the application of progressivism and liberalism, the human race would be localized and inbred, likely still living in caves, serving tribal kings, wearing animal skins and going on raids to kill other men and steal women and food.

So the hyperconservative hates and distrusts everyone and everything, the hyperliberal loves and trusts everyone and everything. The former gets you Hitler, Nazis, fascists, dictators, Catholic States, Islamic States, Inquisitions, Crusades, bigots, religious freedom holy warrior zealots and the current Republican party. The latter gets you taken advantage of, cheated, infected, invaded and overrun. The challenge we face as a species is, frankly, how to best use science and education to marginalize and effectively muzzle the counterproductive hypers on both sides and to identify the best path forward for the most people.

And to get us back to the topic of this book, how does all this psychology and the references to politics relate to religion? The relevance comes from the fact that liberalism and conservatism are not merely political terms confined to a single area of discourse. They describe how humans think and act in the real world in all areas.

Hyperliberal religious beliefs get you Jesus and the penniless hippie healer ethos that he actually preached, the ultrapacifistic faith of Mohandas Gandhi and the nonviolent civil rights campaign of Martin Luther King Jr. Hyperliberal beliefs get you the peaceful Jains and monkish lifestyles in which people burn themselves alive rather than burning other people.

Hyperconservative and even moderately conservative religious beliefs get you the oppressive and evil history of Catholicism, the Inquisition, the Crusades, the Holocaust, slavery, bigotry, suicide bomb mass murders, passenger planes flying into skyscrapers, American antihuman evangelism and people burning other people alive. That one can use the same exact holy books, and in some cases the exact same words, to justify hyperliberal or hyperconservative actions indicates the original and permanent fault in those holy books, which is expressed in those who choose to follow those words to do evil.

Hyperconservatives in America and elsewhere have twisted the teachings of the biblical hyperliberal peacenik Jesus into aggressive campaigns of mean-spirited bigotry, in which only a chosen few supposedly have a mandate from Jesus to own guns, subvert secular law and oppress or persecute certain groups of people. "Jesus wept", indeed.

Jesus would do far more than weep if he were to ever come back and see how his message, indeed his direct orders, had been bastardized and Satanized. If there ever is a Second Coming in America it won't start off with masses of people floating off into the sky toward some magical Candyland, it will start with much flogging and whipping of many millions of the publicly, immorally and hypocritically faithful by the divine and untiring hand of Jesus the Enforcer.

History has shown that religions have almost invariably been used as the single most powerful justification for xenophobia, tribalism, persecution and murder, and for generally bringing out the worst in people.

Christianity is an absolutely perfect example of how this is done in practice. The biblical Jesus character is actually one of the most liberal in all of human literature. Yet his teachings of compassion and forgiveness, ministering to the poor and castigating or even whipping the rich and money-focused, healing the sick and welcoming those in need are often absent in American Christianity.

Indeed, categorizing Christian groups in America in the 21st century is merely an exercise in sorting the different sects by how many different groups of people each specific sect discriminates against, and to what degree.

Many Christians today can only be partially forgiven their hyperhypocrisy when one understands that almost none of them can be bothered to actually read the words of the god they allegedly worship so devotedly. Indeed, why aren't all people who actually follow the teachings of Jesus actually flaming liberals in real life, from their words to their deeds to their votes? Because hypocrisy, xenophobia and greed trump all else, even the actual words of the holy book these people claim as their guide to life.

Religion can never be an arbiter of a truly decent and humanistic morality, and it has proven this for thousands of years in nearly every country on earth. To prove this all one need do is look at how many Jains, Buddhists and Quakers there are, as opposed to how many Christians and Muslims there are. The first three are peaceful and pacifistic groups; the latter have been hyperaggressively hyperconservative, xenophobic, exclusionary and have killed and persecuted far too many people to count. Jain fundamentalists have never fought a single war or killed a single person.

Extremely devout and faithfully fundamentalist Christians and Muslims have killed people wherever they have gone, often in massive numbers and in ways too horrible to describe, all while following the very specific orders and teachings of their holy books and their faith and in defense of these faiths.

One really need no more evidence than this to reject any claim by any of the monotheisms that they have any contributing role in any conversation on morality, now and forever. They have written the books on grotesque immorality, both literally in their holy books and in practice with the blood of their victims.

How does all this happen? How can a person claim to be a decent person when his acts are truly vicious and horrible? How can a person claim that his or her holy book is good and moral, when it is the exact opposite? For that matter, how can a person deny the findings of science when the evidence is ubiquitous, cross-disciplinary and overwhelming?

For this we must refer back to the opening paragraphs of this chapter and to more facts of human psychology. Humans have an uncanny ability to fail to think, to think incorrectly, and to only see what they want to see. Over time, objective science has catalogued a large number of the flaws in our thinking.

Table 1 below is a partial list of human cognitive biases and logic flaws that sabotage reason and critical thinking. While it is somewhat lengthy, this list is nevertheless very far from all-inclusive and does not even include phobias.

Note that every human alive currently operates while under the spell of some number of these traps, and while we theoretically have the capability to discern them in ourselves, we almost always do not. We can't even do a good job of perceiving them in others.

Now that these flaws and biases are known to science, we now have the ability to be told that we tend to fall victim to certain biases, and to understand this and change our behaviors accordingly, but again this almost never happens in practice. As we continue to gain more knowledge of human psychology, we continue to gain a clearer understanding of why people think, say and do what they do.

The first two columns in the table list the bias or logic flaw, a brief description of it and its relevance to religion. The third and fourth columns reflect the degree to which these cognitive traps first enable the invention of and then the persistence of religious or supernatural beliefs, ranked on a scale from 1 to 5 with 5 being the most powerful and 1 the least powerful.

As you read through the table, think about each bias or flaw and how it would be expressed by a person writing the Old Testament, the New Testament or indeed anything written two thousand or more years ago. Think of the perspective of that person, wholly ignorant of the actual facts of astronomy, meteorology, chemistry, physics, geology, biology, psychology, sociology and the other sciences.

In other words, think of how unbelievably ignorant you would be on these subjects if you had been raised by ancient desert goatherders. Also think of how our biases can be reduced by the vast amounts of knowledge which even a poor high school education and life in a First World society tries to put into a person's head today.

Gods are no longer necessary to explain lightning, thunder, earthquakes and illness, and they certainly aren't needed to explain cars, televisions, cell phones, waves, mountains, the origin of life or the origin of the universe. But explanations from not even that long ago for even the most currently obvious and simple problems defaulted to a divinity due to our near-total scientific ignorance.

And this ignorance is the starting point for every single new human. Babies are born with no knowledge of the world and each person has to accrue massive amounts of knowledge in order to reach a properly informed perspective.

Think about how easy it was for people whose world was absolutely filled with imaginary gods and goddesses to attribute everything to those gods, to have his or her thinking conform to a mindset which considered gods as a natural and accepted fact, and thus how imprisoned these people and their society were by religion and religious beliefs.

As you read the table below from that perspective, see just how easily the human brain can first be anchored to supernatural beliefs, then to confirm them, and how hard these ideas are to kill once developed and established. Think about today's believers and how different people can be equally devout followers of religions invented in antiquity such as Judaism and Christianity or of recently invented ones such as Scientology or Mormonism.

*Table 1. Listing of cognitive biases, their religious implications and how they contribute both to the invention of and the persistence of religion in the human mind and society.*

| |
|---|
| **Anchoring.**[39] Over-reliance on one bit of data about a subject, often the first bit of data one hears, while making a decision. The first bit "anchors" thinking on that subject.<br>**Religious Implication.** When young people are indoctrinated into religion, it very often stays with them for life and shapes a great deal of their thinking.<br>**Enables Invention: 1**<br>**Enables Persistence: 4** |
| **Anthropocentric Thinking.**[40] "...the belief that human beings are the most significant entity of the universe. Anthropocentrism interprets or regards the world in terms of human values and experiences."<br>**Religious Implication.** This strengthens the tendency to adhere to anthropomorphic religions like the three monotheisms.<br>**Enables Invention: 1**<br>**Enables Persistence: 3** |
| **Anthropomorphism.**[41] "...the interpretation of nonhuman things or events in terms of human characteristics, as when one...hears human voices in the wind."<br>**Religious Implication.** As above, this strengthens the tendency to adhere to an anthropomorphic religion.<br>**Enables Invention: 1**<br>**Enables Persistence: 3** |

**Attentional Bias.**[42] When thoughts affect perceptions. If a person is thinking of the color red, his perception tends to incorporate redness.
**Religious Implication.** As with anchoring, attentional bias in a believer keeps focus on the faith and its claims and far away from facts, logic and reason.
**Enables Invention: 1**
**Enables Persistence: 3**

**Authority Bias.**[43] The inclination to "…attribute greater accuracy to the opinion of an authority figure (unrelated to its content) and be more influenced by that opinion."
**Religious Implication.** The faithful generally follow the words of their theologians and clergy, regardless of any resulting conflicts with facts, reason and humanism. This is due to the strong "follower" characteristic in many believers.
**Enables Invention: 1**
**Enables Persistence: 3**

**Availability Cascade.**[44] The process by which an idea grows in acceptance and popularity through increased repetition.
**Religious Implication.** Religions rely on the unceasing repetition of the same words and messages, multiple times weekly or even daily. Thus they become lodged in the brain even if they are untrue. As Mark Twain said[45], "History is strewn thick with evidence that a truth is not hard to kill, but a lie, well told, is immortal."
**Enables Invention: 1**
**Enables Persistence: 4**

**Backfire Effect.**[46] When a person responds to evidence disproving a strongly held belief by ***increasing*** the strength of his belief and rejecting unwelcome evidence. It is probably the most damning example of human illogic and perversity, that a person can unashamedly continue to deny facts that are literally undeniable.

**Religious Implication.** Many people resist disconfirming data with whatever force is required to maintain their beliefs. Many others, in the drive to permanently block this information, ramp up their devotion to their beliefs. Thus the faith becomes irrevocably entrenched and many of these people begin to or increase their current efforts to convert others to their beliefs. This is one of the main reasons why unjustified faith is still with us and why the most insecure of the faithful can't keep their religion to themselves.

**Enables Invention: 1**

**Enables Persistence: 5+**

**Bandwagon Effect.**[47] The inclination to think, say or do things because large numbers of other people do.

**Religious Implication.** Sheeple follow fads, and religions are a species of fad. Once a fad or religion grows, herd dynamics and the bandwagon effect combine to keep people in the religion.

**Enables Invention: 1**

**Enables Persistence: 3**

**Belief Bias.**[48] The inclination to evaluate an argument or position based on how closely it aligns with the person's beliefs.
**Religious Implication.** This is why believers reject disconfirming evidence. They don't like how it affects their beliefs, or how it makes them feel, so they reject even incontrovertible evidence.
**Enables Invention: 1**
**Enables Persistence: 5**

**Bias Blind Spot.**[49] The inclination for a person to see bias in other people but not to see bias in oneself
**Religious Implication.** Religions are founded on faith, not facts, and since religions have no self-corrective mechanisms, the blind spot lives on
**Enables Invention: 1**
**Enables Persistence: 4**

**Choice-Supportive Bias.**[50] The inclination to justify one's choices after the decision has been made.
**Religious Implication.** Particularly in religion, which has so many unprovable and unjustifiable claims, the strong desire to avoid embarrassment and criticism and to reassure oneself that joining a faith was the right decision prevents objective assessment of the decision.
**Enables Invention: 1**
**Enables Persistence: 3**

| |
|---|
| **Clustering Illusion.**[51] The inclination to believe that small streaks in data represent a pattern when they actually do not.<br>**Religious Implication.** The religious mind is tuned, trained and encouraged to see patterns that do not exist or to interpret patterns in a way supportive of the faith. It thus perceives patterns which are thought to support the faith, despite the lack of evidence or reason for this position.<br>**Enables Invention: 1**<br>**Enables Persistence: 4** |
| **Confirmation Bias.**[52] The inclination to seek, interpret and preferentially believe data that is believed to support one's existing beliefs.<br>**Religious Implication.** Believers tend to only accept data they think confirms their beliefs, thus failing twice. Failure 1 is to reject disconfirming evidence, and Failure 2 is to twist facts as far out of shape as is needed to rationalize the beliefs.<br>**Enables Invention: 1**<br>**Enables Persistence: 5** |
| **Curse of Knowledge.**[53] The inclination for a more informed person to fail to consider issues from the perspective of the less-informed. This can lead to callousness, insensitivity and an inability to communicate.<br>**Religious Implication.** Religious people are often denigrated as weak or ignorant, and most are, but merely saying so and not taking appropriate action is not being constructive. People are believers for many reasons, and each must be addressed in a way understandable to the believer, so that the path to reality is clear.<br>**Does not enable *formation* of religion**<br>**Enables Persistence: 1** |

| |
|---|
| **Dunning–Kruger Effect.**[54] The inclination for people with low ability, competence or knowledge to fail to see these characteristics in themselves.<br>**Religious Implication.** Everyone thinks he or she is smart and knowledgeable even though many are not, and these people are often the most aggressive in using their feelings as justification for their thoughts and actions. Forrest Gump said it best with "Stupid is as stupid does".<br>**Enables Invention: 1**<br>**Enables Persistence: 4** |
| **Egocentric Bias.**[55] The inclination to think higher of one's abilities and opinions than is warranted.<br>**Religious Implication.** The size and reach of the ego of the true believer, who feels he knows what his god wants him to think and do, have no limits. Thus we see the insufferable egotism of those who claim to know the mind of their god and think the rest of humanity is just a bunch of misinformed people who need to be converted to the faith or left to wallow in their "unsaved" misery.<br>**Enables Invention: 1**<br>**Enables Persistence: 2** |
| **False Consensus Effect.**[56] The inclination for a person to believe his personal thoughts, opinions and beliefs are more widely held than they are. It is self-deception.<br>**Religious Implication.** This very dangerous fallacy reduces individual restraint. When a person thinks everyone else thinks like he does, he often acts on those thoughts, and with religion this invariably leads to antihuman behavior.<br>**Enables Invention: 1**<br>**Enables Persistence: 3** |

**Forer Effect or Barnum Effect.**[57] "…when individuals believe that personality descriptions apply specifically to them (more so than to other people), despite the fact that the description is actually filled with information that applies to everyone. The effect means that people are gullible because they think the information is about them only when in fact the information is generic. The Barnum Effect came from the phrase often attributed (perhaps falsely) to showman P. T. Barnum that a "sucker" is born every minute."

**Religious Implication.** Like the other cognitive traps in which people imagine patterns that aren't there, people make connections that are not there to justify their religious beliefs. Just like when 100 random people can all believe the exact same horoscope is talking about them personally, believers read about things in their books that they falsely perceive to be "real" at some point in history or their lives.

**Enables Invention: 1**
**Enables Persistence: 4**

**<u>Framing Effect.</u>**[58] When people make different choices resulting from the same information, due to a difference in how the information is presented.

**Religious Implication.** The Jesus story is said to be "good news", as he was allegedly sent here to die heroically for humanity's sins, and Catholics thus celebrate the Eucharist. It can be restated as 'the creator of the universe sent his only son to be tortured to death so the sins of the people which the creator had created with these sins built into them could be forgiven, and to celebrate this event people now believe they actually eat and drink the reincarnated zombie body and blood of this murdered half-breed Jewish preacher in magical ceremonies performed by celibate males.' Same information, different presentation, much different processing and response in the reader.

**Enables Invention: 1**
**Enables Persistence: 3**

**Fundamental Attribution Error.**[59] This is "…the tendency people have to overemphasize personal characteristics and ignore situational factors in judging others' behavior. Because of the fundamental attribution error, we tend to believe that others do bad things because they are bad people. We're inclined to ignore situational factors…"

**Religious Implication.** The faithful are notoriously judgmental, and rely on very simple platitudes such as "sin" or the teaching of ancient mythology over the understanding reached through complex analysis of the realities of psychology, sociology, anthropology and biology. Such unscientific and ignorant behavior is antithetical to a civil society.

**Enables Invention: 1**

**Enables Persistence: 4**

---

**Hindsight Bias.**[60] "…the tendency, upon learning an outcome of an event—such as an experiment, a sporting event, a military decision, or a political election—to overestimate one's ability to have foreseen the outcome. It is colloquially known as the "I knew it all along phenomenon."

**Religious Implication.** The religious, being primed to praise their god for good outcomes and chalk bad ones up to his inscrutable will or to dirty rotten sinners, are similarly primed to say they knew things were going to happen as they did, after the fact and despite what they may have previously said.

**Does not enable *formation* of religion**

**Enables Persistence: 3**

**Illusion of External Agency.**[61] The inclination for people to think that outside agents such as a god influence actions they actually take of their own accord.

**Religious Implication.** When people imagine their god guided them to do something that was beneficial, they self-reinforce their beliefs and further cement the illusion of divine intervention into many facets of their lives. Despite the lack of evidence for such belief, this self-fulfilling and self-reinforcing cycle digs the believer ever deeper into the faith, making it nearly impossible for facts to intrude or the mind to change.

**Enables Invention: 1**
**Enables Persistence: 5**

**Illusion of Truth Effect.**[62] The inclination of a person to believe the truth of statements he has heard multiple times, even if the statements are completely and intentionally false. In other words, belief that is accrued through simple repetition.

**Religious Implication.** As with the Availability Cascade, even the biggest lies are eventually thought true if they are repeated often enough. Religion is founded on fables repeated endlessly, so believers accept them despite the lack of evidence. Repetition is reinforcement and the adherence to the faith is strengthened. For centuries, Catholics and Christians persecuted and reviled Jews for imaginary behaviors and this eventually led to the Holocaust. Religions incessantly repeat their propaganda and dogma. Actions in the service of dogma and propaganda are always antihuman and destructive.

**Enables Invention: 1**
**Enables Persistence: 4**

| |
|---|
| **<u>Illusory Correlation.</u>**[63] This is incorrectly thinking that two unrelated factors actually are related.<br>**Religious Implication.** Believers link positive traits like morality with their own in-group. This reinforces belief, tribalism and xenophobia due to perceived in-group superiority.<br>**Enables Invention: 1**<br>**Enables Persistence: 3** |
| **<u>Ingroup Bias.</u>**[64] The tendency for people to treat well those who they believe are in their group, and to treat poorly those who they believe are not in their group.<br>**Religious Implication.** Religious divisions are the most destructive, because believers think they actually have divine authority to support their faith group and fight or destroy out-groups. Examples include religious bigotry, religious wars, the Holocaust, Crusades, Inquisitions, persecution of certain groups and countless other examples.<br>**Enables Invention: 1**<br>**Enables Persistence: 4** |
| **<u>Just-World Hypothesis.</u>**[65] The tendency for people to think that life is "orderly, predictable and just", which often leads to victim-blaming when misfortune or tragedy occurs. In other words, the victim deserved it.<br>**Religious Implication.** Believers have proclaimed for millennia that bad things happen to sinners via karma or divine punishment. They use their religion to justify their lack of sympathy for certain victims, especially if the victim was engaged in an activity the faithful consider sinful.<br>**Enables Invention: 1**<br>**Enables Persistence: 3** |

**Law Of The Instrument.**[66] The tendency to use a familiar method or tool to accomplish a wide variety of tasks. The most common cited example of this is the phrase "If all you have is a hammer, everything looks like a nail."
**Religious Implication.** This also applies to mental heuristics, so a religious person will see and try to explain the world through his own personal religious lens. This is why the religious ignorance and dismissals of religion-disproving science are so pernicious and destructive.
**Enables Invention: 1**
**Enables Persistence: 4**

**Mere Exposure Effect.**[67] The tendency to like things more than would be expected simply due to being familiar with them.
**Religious Implication.** Religions are replete with rituals, ceremonies, services and other events, often in the same place and with the same people. The frequency of exposure strengthens the belief in religion and its teachings, no matter how unbelievable or immoral, and the clergy know it.
**Enables Invention: 1**
**Enables Persistence: 4**

**Naïve Realism.**[68] A person's belief that he sees reality as it really is, while those who disagree with him do not see reality as it really is because the other person is believed to be biased, ignorant or irrational.

**Religious Implication.** Believers think theirs is the One True Faith, and that nonbelievers and believers in other faiths are biased, ignorant or irrational. This is used as justification to continue in the faith and strongly contributes to tribalism. Strong challenges to a group's faith bind members together even more closely, and they will double down on their naïve realism.

**Enables Invention: 1**
**Enables Persistence: 3**

**Ostrich Effect.**[69] Ignoring obviously negative information.

**Religious Implication.** There are far too many examples of this in religion to count much less record in this book. Catholicism has a history written in blood and murder, yet believers blithely ignore this history and ignore painful truths. The global child rape scandal which isn't causing people to leave Catholicism is just one example. All religious believers, especially Christians, have an uncanny ability to ignore anything that puts their faith in a bad light or disproves their religion.

**Enables Invention: 1**
**Enables Persistence: 4**

**Overconfidence Effect.**[70] Similar to the Dunning-Kruger effect, this is the undeserved confidence people show in their answers to questions. People almost invariably have high confidence in their beliefs and answers, despite these answers being very wrong in many cases, with error rates as high as 40% even when the answer is given with extremely high confidence.
**Religious Implication.** Overconfidence is common in humans, and science is the best method we have to eliminate it. Religious beliefs are reinforced by repetition, by groupthink and by each believer's internal need for the belief to be true, thus believers cling to undeserved overconfidence. This is obvious when asking believers for evidence. Even though they cannot give any, they astoundingly maintain great confidence even in the minute details of beliefs for which there is neither justification nor evidence. And when a person imagines that his god has told him something, there is no changing his mind.

**Enables Invention: 1**
**Enables Persistence: 4**

**Pareidolia.**[71] "The tendency to perceive a specific, often meaningful image in a random or ambiguous visual pattern."

**Religious Implication.** People imagine the faces of Yahweh, Jesus or his mother or any of their countless saints and revered people in clouds, food, trees and smoke. In his book *The Believing Brain*[72], Michael Shermer discusses how people hear imaginary messages and perceive nonexistent patterns in static and noise. This ties directly to religious belief. Christians and Muslims never see or hear from Zeus or Marduk when they hallucinate or pretend to commune with the divine, proving how extremely provincial a believer is when imagining things where none exist.

**Enables Invention: 2**
**Enables Persistence: 4**

**Post Hoc, Ergo Propter Hoc.**[73] Latin phrase describing a very common fallacy. It means "After this, therefore because of this: because an event occurred first, it must have caused this later event —used to describe a fallacious argument." An example is that if my phone rings or I yawn after a dog barks, the barking did not cause my phone to ring or me to yawn. The barking happened to precede the ringing or the yawn. No gods were involved.

**Religious Implication.** Religions, being founded on unprovable claims, are the human activity most susceptible to logical error. This particular error is very powerful and is very common in religion. Believers are primed to believe their gods intervene in the world, and due to the attentional bias, the availability cascade and the confirmation bias, they ignore bad events and always notice good ones in a way they like. Thus if they pray for X and it happens, they think their personal prayer to their personal god was the cause. Billions of people currently suffer from this insane delusion and its power is greatly magnified in the believer's head because of all the other biases that also destroy critical thinking.

**Enables Invention: 3**
**Enables Persistence: 5**

**Post-Purchase Rationalization.**[74] The inclination to persuade oneself after the fact that a purchase or decision was the right one.
**Religious Implication.** Denial isn't just a river in Egypt, and humans can rationalize absolutely anything. The compulsion to reject the possibility of error in order to maintain one's self-image as a good decision-maker means even horribly bad decisions will be rationalized away. Thus many people will remain in their religion no matter how bad the choice or the religion. That a billion people are still Catholics despite the global rape epidemic and its coverup by their church's leaders proves the power of such rationalization.
**Does not enable _formation_ of religion**
**Enables Persistence: 4**

**Selective Perception.**[75] People see what they want to see and hear what they want to hear. A thousand people can watch an event and each can come away with different interpretations of what happened based on their personal biases.
**Religious Implication.** Like the confirmation bias, believers see the world through the filter of their own particular religion, so a factual understanding of an event is difficult to achieve. This is even for events of which we have full video and audio.[76] The believer often accepts only what confirms his perceptions.
**Enables Invention: 4**
**Enables Persistence: 4**

**Self-Serving Bias.**[77] The inclination to feel more personal contribution to personally experienced successes than failures, and to feel less personal responsibility for those personally experienced failures.

**Religious Implication.** Humans have an unlimited ability to rationalize. Believing good things happen to you because you follow the One True Faith and its god or prophet, and attributing bad things to a devil or evil spirit, is extremely common among believers. Many Christian groups focus almost exclusively on sin, shame and guilt in all its masochistic expressions, but one can be part of these cults and still keep the self-serving part of this bias, along with an unwelcome zeal to spread this belief to others who do not share it.

**Enables Invention: 1**
**Enables Persistence: 5**

**<u>Semmelweis Reflex.</u>**[78] This is a nearly universal and very significant problem not usually referred to by this term. It is "…the reflex-like tendency to reject new evidence or new knowledge because it contradicts established norms, beliefs or paradigms." It is a childish rejection of things which challenge your current beliefs.

**Religious Implication.** Religions center on their ancient holy books, teachings and traditions, and they demand their flocks adhere to them. Every time science shows religions are full of errors, ignorance, immorality and bad ideas, and every time science makes a new discovery that disproves some element of faith, many believers reject the information out of hand. They maintain this rejection until some believer dreams up a way to rationalize the discovery as being part of the god's plan after all and even more proof of the god's existence.

**Enables Invention: 1**
**Enables Persistence: 5**

**Shared Information Bias.**[79] The Wikipedia definition is so good and so relevant to religion that I included the first few sentences here. It says"…the tendency for group members to spend more time and energy discussing information that all members are already familiar with (i.e., shared information), and less time and energy discussing information that only some members are aware of (i.e., unshared information). Harmful consequences related to poor decision-making can arise when the group does not have access to unshared information (hidden profiles) in order to make a well-informed decision."

**Religious Implication.** Religions rely on faith because they can't prove any of their claims. Thus they engage in the endless repetition of the faith's mantras and dogmas in services and between and among members to continue the groupthink and reinforce the in-group tribalism. Information from the outside is not welcome due to the damage it can inflict on dogmas. The herds which comprise the religious flocks do not like to leave their comforting echo chambers and self-congratulatory and self-reinforcing meetings. Believers don't want their superhero gods or their beliefs being exposed to any Kryptonite-like realities. Hosting "why we believe" parties is more their style.

**Enables Invention: 1**

**Enables Persistence: 4**

**Status Quo Bias.**[80] The tendency to like things to stay just as they are, or with very minimal changes. This is very strong in conservatism and the conservative and religious mind.

**Religious Implication.** Religions abhor change, because change would be a massive admission they had been wrong about their god's message. Therefore the books do not change and the teachings change as little as possible and only when under tremendous pressure. And when faiths do change, people often leave, which is highly undesirable to the religion's leaders. Also, once a person has ties to a faith group, the social cost of leaving is almost always unacceptably high. Thus the status quo is very strongly maintained.

**Does not enable *formation* of religion**

**Enables Persistence: 5**

**Subjective Validation.**[81] Wikipedia again has such a good and relevant definition that I used it here. This is a "...cognitive bias by which a person will consider a statement or another piece of information to be correct if it has any personal meaning or significance to them. In other words, a person whose opinion is affected by subjective validation will perceive two unrelated events (i.e., a coincidence) to be related because their personal belief demands that they be related."
**Religious Implication.** This bias maintains dogma and belief in the minds of the faithful. Much like a non sequitur is a wrong conclusion, subjective validation means that a believer is highly likely to attribute a positive event to his god if it will make him feel better to do so. If a negative event befalls someone he does not like or someone who he thinks is a sinner or somehow deserves punishment, this is also taken as proof of the god he worships. This is often seen when believers think their god has sent disaster to a community for some imagined sin by one or more of its residents, be it Sodom and Gomorrah, San Francisco, New York City or New Orleans.
**Enables Invention: 1**
**Enables Persistence: 4**

**Sunk Cost Fallacy.**[82] This is "the idea that a company or organization is more likely to continue with a project if they have already invested a lot of money, time, or effort in it, even when continuing is not the best thing to do." It is deliberately continuing with an action[83] beyond when reason and evidence show that the action should be discontinued, as a result of feeling pressure to continue with the action due to the large amount of resources already invested in the action. Also known as Irrational Escalation.

**Religious Implication.** Religion is full of examples. When a believer is heavily invested in a religion, particularly socially or due to the volume of time and resources placed into the faith for long-time believers, leaving it is seen as less desirable than just staying the course even after grave disillusionment. I again have to note how many believers are still Catholic even after the global child rape scandal and the equally horrific coverup from the very top all the way down. And even after so many people remained Jewish after the six million dead in the Holocaust. There is just no shaking some people out of their unjustified beliefs.

**Enables Invention: 1**
**Enables Persistence: 5**

**Survivorship Bias.**[84] Focusing on the survivors of an action or process and disregarding those which or who did not survive, due to the lack of visibility of the non-survivors after the event or process. **Religious Implication.** Out of many, the most relevant implication is the near total ignorance most people have about every other religion or god humans have invented except the one they are in. Countless religions, sects and cults have formed and died, and there is no reason to believe any of the current ones are any more correct than any of the dead ones. And the horrifically high numbers of those who die from "Acts of God", which are just godless natural disasters, are ignored in favor of the few so-called "miraculous" survivors. This is pathetic and ignorant. A true miracle isn't one person surviving a disaster or plane crash. A true miracle is every single person surviving unhurt, or a plane landing safely despite losing a wing. Such simple and obvious failures of logical and critical thinking strongly enable religious belief.
**Enables Invention: 2**
**Enables Persistence: 4**

The above listing is obviously quite extensive, but again it is very far from complete. Unfortunately, due to our national failure in America to educate ourselves and our children on basic human thinking, most if not all of these fallacies and biases will be new to many people. Anyone who reads through them all will surely have noticed at least some in other people.

One hopes that readers will also be able to see them in themselves, and thus see human thought and interaction in a more informed light. Again, the reason for including them in this book is that many of them contribute directly to the formation then to the persistence of religions and religious ideas.

As a quick summary, the following is a general flow of how these mental traps have enabled the massive accumulation of religions, beliefs and believers we have seen across our world.

Humans have been able to speak for a very long time. One study[85] estimates that language began to form as much as 1.75 million years ago. Whether it started this long ago, or when we evolved into our current level of genetic modernity approximately 200,000[86] years ago, or during a Great Leap in human evolution approximately 50,000 or 60,000 years ago, it is assuredly the case that during much of our evolution we shared most if not all of the basic patterns of both our modern human cognitive capabilities and our biases.

Thus some people were intelligent, some were inquisitive, some were slow of mind, some were gullible and some were skeptical. Once these humans began to speak, they started to discuss things like what caused rains, thunder and lightning, where the world and the life on it came from, what dreams were and what they meant, where babies came from and why so many of these babies and their mothers died. Their discussions and decisions were influenced by their own personal cognitive biases and the events in their short and difficult lives and their tiny social groups.

The brains of humans born 40,000 years ago were much the same as ours are today. If she were transported to the present day, a baby born 40,000 years ago to a Stone Age hunter-gatherer would grow and develop like a baby born today. Our society has accumulated an unbelievably large amount of information but absolutely none of it enters the brain except through a lifetime of slow learning, filtered through the cognitive biases listed above and many others not listed. Every newborn baby's brain is essentially a blank slate except for an onboard operating system, some processing modules and a few instincts. So the hardware and software of today's humans and that of Stone Age humans are largely the same.

The only difference is the information now available for the growing person to take in.

Thus a person born in the year 2400 could be raised in an environment which exactly duplicated the Stone Age, and she would rise only to the level of the environment and the small amount of information passed along from those around her. This person would know nothing about anything at all except flaking stone, making fire and stone axes, shell and bone jewelry, spears, gathering food and whatever cultural rites, games, wars and other behaviors and technologies those around her exhibited.

There would be no Yahweh, no Jesus, no Allah, no Buddha, no Greek or Roman pantheon, no Santa Claus and no Tooth Fairy. What's more, there would be absolutely no way to derive these fantasies and the person would be completely fine without any of them. If the person were raised to believe that giant purple magic rabbits caused thunder, or that it was Stone Age angels doing the Stone Age equivalent of heavenly bowling, the person would believe this.

Due to the scarcity of both knowledge and answers to questions of causality and origins, when a particularly garrulous or charismatic person in the Stone Age imagined a solution to the questions of the origin of life and the earth and proselytized it, many people would likely believe it. This would have happened just as many people in the ancient Middle East believed the local prophets and came to believe in polytheisms and later in the monotheisms of Judaism, Christianity and Islam. The same process occurred in the USA in the mid-1800s with the invention of Mormonism and again in the mid-1950s with the invention of Scientology.

The process is exactly the same. Somebody came up with a fairy tale, other people followed it, then it got bigger and achieved "critical mass" sufficient to keep it alive for some time yet TBD.

So when Og the Stone Age philosopher looked around and wondered how everything began, he saw a world in which only humans did "intelligent" things like shaping stones and making axes and fire. Animals and plants did not do this. Therefore he posited that the heavens, the world and the events in them were the result of some super-powerful human or humanlike entities. Thus the "super-human" entity or entities, some of which were part human and part animal, were likely the first supernatural gods which humans invented to explain the mysteries.

The original stories were passed on around millions of campfires over the years, growing in detail and complexity to try to explain more mysteries and, in many cases and much more importantly, to captivate the audience. Captive audiences result in more attention being directed to the speaker, and quite often also more social status, power and sex, thus this sort of behavior is now a near human universal. Bear in mind that almost everyone at that time, and far, far too many people even unto this day, believed that dreams are some kind of message from or insight into a supernatural realm. Thus nearly everyone was primed for explanations which included imaginary realms.

Later generations of Stone Agers, you could call them campfire televangelists, discovered that they could tell their clanmates that they had heard voices in their heads or seen visions or signs which were communications from the beyond, and that people would actually believe these stories. They soon learned that they could use these assertions of communications with the gods as a justification to tell everyone else in the clan how to behave and also to gain sex, power and influence. Thus formal religions and religious hierarchies were born, normally being run by power-hungry authoritarian sociopaths.

Replace Og and the campfire televangelists with any of the prophets, mountebanks, charlatans and other clergy of any other religion throughout human history and you have described how religions form and are sustained. Original ideas born of ignorance, insanity, cupidity, malevolence, powerlust, a compulsion to be the center of attention, hyperactive imaginations or an unfortunate combination of these unsavory characteristics were accepted by the masses and eventually infected millions then billions of people.

Humans began the process of trying to explain major mysteries with almost zero knowledge of the actual facts of the world and the universe. The only tools they had for this were their five senses and brains that were programmed over millions of years to seek very local, very small-scale and very short-lived patterns in a savannah environment.

This was rather like trying to explain how electricity works and how it can be harnessed using a lightning bolt as your baseline. Only through actual knowledge of physics and chemistry can one start to understand then describe electricity.

It is only since 1859 that humanity took its third, and some would say its most significant, great modern intellectual leap forward. The first of these leaps was the Scientific Revolution which began in the mid-1500s, and the second was the very appropriately named Enlightenment, which began in the mid-1600s. The Scientific Revolution and the Enlightenment overlapped in time and in many intellectual areas for much of the 1700s.

Humanity spent over two million years in a pre-scientific world, then in relatively short succession following a series of climatic and genetic developments, developed a series of increasingly more powerful tools and instruments to collect and process data. The ancient Mediterranean schools of thought and philosophy gave way to the Middle Eastern scientists and thinkers of the Islamic Golden Age which lasted roughly from the 700s to the 1300s. The baton of scientific progress then passed to the growing population of scientific thinkers in Europe who were trying to come out from under the Inquisitional thumbscrews of the Catholic overlords.

The global rate of accretion of scientific knowledge has been positively asymptotic since the mid-1700s, but natural selection was not widely discussed in detail until after 1859, which on the scale of our existence happened mere seconds ago. Genetic engineering, quantum theory and the Big Bang did not arrive until the 1900s. Much of our detailed knowledge of the physical universe has only been with us for only a microscopic fraction of the existence of our species, and not nearly long enough to displace the stultifying inertia of centuries of religious indoctrination.

This accumulation of knowledge has progressed unbelievably quickly, but the human brain has not progressed or evolved. The people who wrote the holy books of Judaism, Christianity and Islam are the products of minds from a highly ignorant pre-scientific age which shared not only each and every human cognitive bias but also a near total ignorance on a huge range of scientific disciplines. They were able to generate much in the way of philosophy, literature, art and some math, but science and the realities which it unlocks were far in their future.

The difference between the 21st century and every single year in biblical times is the almost unimaginable amount of knowledge we now have compared to then, including knowledge of the sciences and of our biases and logical fallacies.

Think about it this way. Compared to the level of knowledge we have today, the average person living in the ancient Middle East during biblical times knew about as much about physics, chemistry, psychology and other sciences as did our flint-knapping ancestors of two million years ago. And we should treat their books accordingly.

A person reading those ancient fossilized holy books today can only use them in a supernatural or religious sense if he does not know about or rejects science, does not know about or claims total immunity from psychological biases, and does not know of the thousands of other religions or does know about them but rejects them all in favor of one particular preferred faith. These are all utterly ridiculous and unjustifiable positions.

Getting back to the brain itself, there are more considerations about it which we must address before closing out this discussion on the first source of our beliefs - the physical brain itself.

The brain functions in accordance with the interplay of chemicals and electricity. Not every brain has the exact same chemistry, the exact same amount of different neurotransmitters or the exact same efficiency or speed in processing them. Variations in the amounts, availability and persistence of chemicals like those neurotransmitters have significant effects on the function of the organic machine, and therefore on the thoughts which the machine produces.

The effects of chemicals like alcohol, uppers, downers and hallucinogens are examples of what happens when brain chemistry temporarily changes. Minor differences in the levels of chemicals naturally produced in each brain drive it to function in a manner slightly different from other brains.

While not quite in the same category as hallucinogens, the neurotransmitter dopamine[87] has been shown to be linked to the level of belief in unproven assertions and paranormal thoughts.[88] Higher amounts of dopamine are associated with higher rates of belief in the supernatural and paranormal. This is a very significant finding which has been in the literature for some time now.

One study[89] published in 2015, which involved both genetic analysis and formal psychological experimentation and analysis, was conducted by seven researchers and is titled "Linking unfounded beliefs to genetic dopamine availability". This study noted the following: "Our results indicate that genetic differences in dopaminergic neurotransmission account for inter-individual differences in perceptual inference linked to the formation and maintenance of unfounded beliefs."

In other words, people who naturally produce more dopamine and/or have a greater sensitivity to it are more likely to have and hold "unfounded beliefs" such as those in religions. This has significant implications. Just as a person who has taken in alcohol, cocaine, heroin, morphine, methamphetamine or a hallucinogenic drug experiences a change in his mental state and perception of the world, a person with naturally higher levels of dopamine will perceive the world differently from a person with lower levels.

Research has shown[90] that believers are, in layman's terms, more gullible than skeptics. This is of course no surprise to anyone, but the purpose of science is to confirm or deny hypotheses and so-called "common sense". The threshold for evidence required by the believer is far lower than that required by the skeptic, and this is most pronounced with religious or mystical claims.

Belief in the claims of religion without absolutely massive amounts of evidence is quite literally impossible for skeptics, just as it is impossible for them to believe that a fat and permanently old man rides a magic sled through the sky pulled by trained magic ruminants and carrying a world-full of trinkets pulled from a magic backpack tesseract which are magically delivered in one night to millions of young recipients in quite glaring violation of the laws of physics.

Religion goes many orders of magnitude down the road of fantasy in its insupportable claims than does the story of the jolly old elf, and both are equally imaginary. There is actually far more evidence for Santa Claus than for any god, because on Christmas morning there really are actual presents under the tree. No god has ever delivered anything to humanity.

The implications of the above are very interesting. Two brains could be exactly alike except one produces more dopamine or processes it differently, and due to this difference it is possible that one could be a devout religionist or spiritualist while the other could be a skeptical materialist atheist. Perhaps one day we will be able to implant a dopamine suppressor or regulator in the body, and people brought to a normal balance will stop believing that Elvis and Jesus are coming back any day now, that the voices in their heads are divine communications, that they hear messages in what is really just static, that they see faces of gods in fog and smoke, that the government is maintaining a fleet of UFOs at Area 51 along with the remains of the crew, and that gods exist.

**Religion really is just all in the believers' heads.**

A related issue but separate from the cognitive biases and the influence of dopamine is our pattern-detection module. It is one of the strongest modules in the human brain and is both a great asset and one of our greatest weaknesses. Matthew Hutson describes this in a wonderful article[91] in *Psychology Today*, and it has also been extremely well-discussed by Michael Shermer and others. Shermer's term "agenticity" sums it up perfectly. Human brains seek patterns, and the brain's pattern-seeking module runs on programming that operates much the same way today as it did many hundreds of thousands and likely millions of years ago, when we knew absolutely nothing of the natural world, much less our current massively artificial world.

Thus what would almost have been called a correct and "common sense" interpretation of an event for the vast majority of human history often results in absolutely massive errors when it is applied today.

The best way to describe this is to note how humans tend to anthropomorphize so much in the world and assign agency or conscious intent behind the things that happen. The ancients used to think gods either sent or withheld the rain, as apparently many American Christians do today, especially in Texas[92].

This was the case with lightning, thunder, floods, disease and anything else particularly bad or good that happened around or happened to ancient humans. The example Shermer and others use is a rustle in some tall grass heard by two of our ancestors while they were out walking together.

One of the walkers assigns agency to the rustle, assumes it is a man-eating beast and gets the heck out of the area. If the rustle really was a predator this person will survive and have children who will share his "all events have an intentional cause" approach to life. The other hunter-gather assumes the rustle was caused by the wind, watches the other guy beat feet, hangs around to eat a berry and within moments becomes dinner for the beast. This guy has removed himself from the gene pool and will no longer pass on his genes for *not* seeing everything as having intentional agency.

This assignment of intent and agency to the things that happened in the world was an evolutionary adaptation that worked well in our science-free distant past, because rustles sometimes did mean hungry or dangerous beasts. But the same mindset that assigned agency to rustles in the brush also assigns agency to thunder, lightning, rain, disease and other events, and this is a complete and utter failure to understand reality.

The difference between the choices is that guessing wrong about the beast gets you killed. Thus assigning agency is extremely strongly reinforced and "Darwinically" becomes part of the instinctive System 1. But assigning agency to the moon or the rain is neutral. There is no negative result from assigning agency to natural events because unless you form a religion and start sacrificing kids or waging war because of your gods, wrongly believing the gods bring the rain doesn't impact your survival one way or the other.

Thus the net evolutionarily successful strategy of assigning intentional agency became part of our ancestral System 1 and developed over time to assign conscious agency and intent behind things that happen in our world. As a result massive numbers of people today make a tragically widespread logic and category error by assigning agency to a fictional god to try to explain far too much.

In a world full of scientific knowledge this agency-assigning module is an artifact contained in our instinctive System 1 which often gives very wrong answers, no matter how right they feel. In other words, we cannot trust these instincts and the instinctive feelings they produce, because they are flat wrong.

The information we have has come in far faster than our brains could evolve, so this module has not changed and is like a stone ax brought to a modern battlefield. It can work in one or two situations but by and large it is ineffective and counterproductive, because stone ax wielders don't do well against tanks and machine guns. Humans did not lose the module as our population slowly grew then grouped into cultures. Today's believers are slaves to their hyperactive and misguided Stone Age instincts and they still attribute natural occurrences to their god of choice.

Making all this much worse is the fact that the advances of technology have outstripped the rest of the process of human brain evolution by millions of years of relative change. Human brains evolve at a very slow rate, and it has taken about seven million years[93] for our brains to arrive at their present state of evolution from the time we split off from our common ancestor with chimpanzees.

Computer technology, on the other hand, has enabled an approximate doubling of the number of transistors which can be placed on an integrated circuit every year or two that passes, according to Moore's Law.[94] This has led to computer technology moving from the 1940s ENIAC which was the size of a room and weighed as much as a small fleet of cars, to today's computing devices which are light years more powerful and can be worn like a wristwatch. In terms of speed, the ENIAC plodded along like a comatose snail trying to pull a battleship up a steep hill, doing calculations at a rate of about 500 FLOPS (Floating Point Operations Per Second). Today's supercomputers, made barely half a century after ENIAC, whiz along at more than a thousand trillion[95] FLOPS, and new speed records are set every few years.

Supercomputers and technological advancements in literally every scientific discipline have enabled humanity to reach places and gain information in ways the human brain and sensory input systems will never reach on their own, no matter how many millions of years of evolution and natural selection we have in front of us.

So the human brain is a powerful organic machine when compared to the brains of other species, but it is highly susceptible to developmental and environmental damage and influence. It has built-in onboard programming and processing challenges. Our so-called "common sense" is often extremely wrong although it feels right to us.

Thus until such time as humans can genetically engineer a more powerful brain without such cognitive biases, or we can somehow hybridize the human brain with powerful computers to do this, we will be using very ancient software to filter and process information. And we will continue to have factually incorrect but deeply and sincerely held opinions for which we have no justification. This is the reality of our situation and we have to understand and accept it.

Thus we have to continually and deliberately keep mere uninformed opinions on religious matters separate from any actions taken by individuals and societies until the opinions have passed rigorous logical and scientific analysis and testing and have matured from ideas to hypotheses to theories. Processes like the scientific method, peer review and replication of study results must always be the sole final arbiter of rational decision making and public policy. Beliefs and opinions are often not just hot air, they are poisonous hot air.

This first section has been somewhat long and involved, but it was a necessary step in understanding the sources of our beliefs and opinions and how and why they are formed and clung to so tightly. Now that we have described the organic machine in our heads and how it operates when left to its own devices, we must describe the remaining three sources of beliefs, which result from changes or damage to the brain and the inputs it receives from the world.

## Belief Source 2: Physical Changes Or Damage To The Brain From Physical Causes

While the brain's structure does drive beliefs, there is far more to the story of those beliefs than just the effects of genes and how they lead the brain to develop and to generate chemical balances or imbalances. The conditions within the mother's womb play a major role in fetal brain development. They can be products of the mother's acts of omission or commission or things that happen to her that are outside her control. We know that eating well[96], taking prenatal vitamins, exercising, reducing anxiety[97] and stress[98], not smoking[99] or drinking or taking drugs or indeed ingesting anything but filtered water, vitamins and organic food free of pesticides[100] and other chemicals are all activities which strongly contribute to optimal outcomes for the growing fetus and specifically for fetal brain development.

Significant damage to the fetal brain can occur if the mother drinks alcohol while pregnant; this is called fetal alcohol syndrome.[101] If the mother does not take in enough folic acid, this can led to an almost uniformly fatal brain defect called anencephaly.[102] Many other conditions and situations result from improper gestational nutrition and environment.

Many of the impacts to the gestational environment are under the mother's control, while others are not. Food and water free of pesticides and other contaminants may not be available or affordable. The mother's relationship with her family may be extremely stressful. The mother may be forced to endure the stresses of war, religious or other oppression and economic or other distress.

Women, merely because they are female, suffer far more stress and oppression than males do at the hands of religion, society and invading armies, to say nothing of general stressors like pay inequity or the disappearance or lack of financial or parental support from the baby's biological father. The woman may herself be mentally damaged and unable to take proper care of herself, let alone her developing fetus. The mother may contract a disease which affects the development of the fetal brain, such as the Zika[103] virus which can cause a range of quite damaging problems[104] such as microcephaly.

All of these conditions impact the development of the brain *in utero*, and these same threats to proper development continue long after birth. After they are born, too many children are subjected to secondhand smoke, given alcohol, fed poor quality food or not fed at all, left unsupervised and are able to ingest poisons or damaging substances, and/or are exposed to many other similar dangers.

Once out of the womb and in the world as an independent organism, they face new threats to their bodies and brains, such as physical abuse, physical neglect, the potential for accidental injury and a host of other possible impacts to proper brain development. There are over twenty years from birth to the major completion of brain development, and the brain is vulnerable for this entire time.

Damage or changes to the physical structure of the brain can and often do affect how and what people think and believe. Some may be temporary but others are permanent, and it is possible that one single experience may permanently change a person's belief system. Due to the general lack of knowledge about science, psychology and the neurobiopsychological functioning of the brain, this can lead to believing in some religion or god. Consider the following examples of how physical changes to the brain can change thoughts and beliefs.

1. **Phineas Gage.**[105] This extremely well-known case is taught even to most high school psychology students and has been for many years, for obvious reasons. Gage[106] had a large metal rod shot clean through his head in a railroad construction accident in Vermont in 1848, but remarkably, though *not* in any way "miraculously", survived. He experienced

significant changes[107] in his personality, as one would expect from such a serious brain injury. I recommend readers take a few minutes to review this absolutely fascinating case for themselves.

2. **Fetal alcohol syndrome.** This and other conditions occur in children when their pregnant mothers consciously use or unknowingly encounter substances which cause heartbreakingly deleterious effects on the developing brain and the child's behavior and cognitive ability.

3. **Lobotomy.** This act of deliberate brain damage is known to change personality and cognition.

4. **Alcohol-induced inebriation.** Everybody knows how behaviors can change when people drink. You can start with three biological siblings from the same biological parents, and the siblings can have three different views on religion. One is a strong believer in one of the tens of thousands of Christian denominations. Another is a devout atheist and skeptic. A third is a New Ager who carries "magic" stones in his pockets and wears pyramid T-shirts. All these beliefs are irreconcilable and held in different brains, yet when they all go out to a bar to drink, each of their brains will be affected by the alcohol they consume. They won't be equally affected, but the change in the brain's chemical environment will manifest itself in each sibling. One sibling might become depressed and sulk in his cups at the bar. His sister might become overly friendly with fellow patrons while his brother might remove his clothing and dance on

a table with a lampshade on his head while swinging his underwear around like a lasso and saying quite inappropriate things.

All would likely say if asked at the time that there was no problem with their current behavior, but while they did act unusually when compared to their normal behavior, no permanent effects are likely to be manifested after the alcohol wears off since they resulted from the ethanol's transient chemical effects.

5. **Traumatic brain injuries (TBI).** Such injuries are experienced by soldiers in combat, football players and others who suffer severe head trauma such as babies with shaken baby syndrome.

6. **Brain damage** due to any number of other causes, such as strokes, meningitis, anoxia and exposure to lead or other dangerous substances.

7. **Hallucinations.** According to Merriam-Webster, a hallucination[108] is "*1a : perception of objects with no reality usually arising from disorder of the nervous system or in response to drugs (such as LSD)…b : the object so perceived ·wasn't sure if the rain was real or a hallucination; and 2 : an unfounded or mistaken impression or notion : delusion.*" The category of hallucinations is remarkably broad and it appears to be one of the most significant contributors to sudden adoption of deep religious beliefs, particularly when combined with genetic predisposition and social programming.

This is because some people give supernatural meaning both to their own individual hallucinations and those of others such as self-appointed prophets or "holy men". Note that due to the confirmation and other biases, hallucinations are felt to be supportive of a person's own faith or the one he is most familiar with, not some unknown or different religion or god.

Hallucinations are auditory and/or visual manifestations of temporary changes in the brain, and while they may seem extremely real to the person experiencing them, hallucinations are merely the brain's failure to properly represent objective reality. Hallucinations of weird colors, shapes and patterns are as much a breakdown in normal processing and perception generation as are hallucinations of gods and the supernatural.

The brain presents our conscious and subconscious selves with a full-sense representation of the world, and hallucinations are distortions of this representation. The brain and only the brain is the source of all thought and experience and it has no tie to any fictional netherworld. Hallucinations are relatively common, affecting about five percent[109] of the general population. Many things cause hallucinations, as noted below.

a. **Psychosis** (schizophrenia/schizoaffective disorder). Research into psychotic hallucinations by Ryan Teeple, Jason Caplan and Theodore Stern[110] noted that "Visual hallucinations in those with schizophrenia tend to involve vivid scenes with family members, religious figures, and animals." This work is the source for "a" through "k" in this section.

b. **Delirium**, which may itself be caused by[111] "…myriad medical conditions, metabolic disturbances, infections, drug effects, and intracranial processes" or from "… alcohol withdrawal (i.e., delirium tremens) or stimulant intoxication (e.g., with cocaine or methamphetamine)…"

c. **Dementia**, specifically Lewy body dementia[112] and dementia due to Parkinson's disease and posterior cortical atrophy.

d. **Charles Bonnet syndrome**, which causes people with failing vision[113] to hallucinate.

e. **Anton's syndrome**, which is an extremely odd and fortunately rare condition. Affected people have become blind due to damage to their occipital lobe, but refuse to believe[114] they are now blind and they act as if they can still see.

f. **Migraines**.

g. **Peduncular hallucinosis**. These are visual hallucinations caused by irregularities[115] in the midbrain.

h. **Sleep disturbances.**

i. **Tumors** on the optic line and in the temporal lobe.

j. **Creutzfeldt-Jakob disease**.

k. **Diabetes / hyperglycemia.**[116]

l. **Lack of sleep**[117] **and extreme fatigue**. Note that these are different conditions from the sleep disturbances noted above. This is another extremely interesting topic worthy of further separate reading. In an article[118] in Scientific American in 2010, Michael Shermer discussed the variant of hallucination he terms "The Sensed-Presence Effect", and what John Geiger calls "The Third Man Factor"[119] in his book[120] with that title. People who experience some combination of extreme fatigue, extreme stress and lack of sleep often imagine the nearby presence of an entity despite nothing actually being there. These internally-generated imaginary experiences seem as real to the person enduring them as actual external experiences do to a nonstressed person.

   Many people throughout history have claimed to have had such an experience, and many of them attributed it to a supernatural entity which they worshiped or knew about. This is undoubtedly due to the widespread distribution of religion and the likelihood that someone experiencing this type of hallucination would attribute it to a known god or spirit rather than their next door neighbor or third grade math teacher. Indeed, it is the rare neighbor or math teacher in the First World today who is widely believed to have supernatural powers.

Shermer describes in great detail the biopsychological conditions which lead to the sensed presence effect, and in a few short paragraphs shows these imagined connections are merely biological byproducts of the human brain. He completely collapses the sail which the faithful have hoisted for centuries in their attempts to prove their line to the divine.

m. **Seizures**. Studies and observation have shown that seizures from temporal lobe epilepsy (TLE) are often strongly interpreted as religious or even "hyper-religious" experiences by the sufferer, with potentially lifelong impacts such as religious behavior and/or conversion. Recent examples and case studies describe this quite well. The researcher VS Ramachandran noted[121] that "When we showed them (TLE patients) religious icons—a Star of David, a cross, or the word "God", the word "Jesus", the word "Mary"—there's a big jolt in the galvanic skin response, showing that something is going on in the temporal lobes, in temporal lobe epilepsy; maybe it's going on in all our temporal lobes. There's a group of neurons that is firing in a normal manner, which makes you more prone to religious belief. And these neurons are hyper active in these people, hence the propensity to religious belief."

This is extremely relevant to the history of Christianity, since many people have speculated that Saul of Tarsus experienced a TLE seizure on his sojourn to Damascus, and that his hallucinations of the divine were nothing more than an epileptic seizure. This proposal is of course dismissed by the faithful but it is infinitely more probable than their assertion that the incident was a communication with the supernatural, and it has quite large implications for the entire corpus of Christianity.

Even beyond the temporal lobes, brain imaging and other data show[122] that seizures affecting a smaller piece of the brain under the temporal lobes called the left anterior insula provide a shotgun blast of effects described as "…an aura of a "sensed presence," a religious emotion, or feelings of euphoria (ecstatic seizures) are characterized by heightened self-awareness."

So your god really is completely in your imagination, your feelings are solely products of the chemicals and electricity in your brain unconnected to any god, and scans and other data can now show this. Yet the faithful dismiss all this science with an airy wave of their holy books.

n. **Near death experiences**[123] (NDEs). I won't go into these very much because massive amounts of work have proven that these are, like the other situations listed above, merely how the brain responds to stress, stimuli and major changes in its environment. Like other hallucinations, they are internally generated sensory manifestations of changes in the brain which include perceived sights and sounds and which often occur similarly between different people. NDEs are merely hallucinations produced by a human brain undergoing unusual circumstances, and they are certainly not supernatural in any way.

o. **Hallucinogenic substances.**[124] There are quite a few of these, and the magnitude of their effects on the user depends on individual body chemistry and the amount of substance used/absorbed. Examples include atropine, dimethyltryptamine (DMT), dopamine agonists[125] (agonists are essentially enhancers), lysergic acid diethylamide (LSD), marijuana[126], MDMA (ecstasy)[127], mescaline[128] (in peyote), phencyclidine (PCP) and psilocybin.

Thus there are countless ways in which a simple physical change in or damage to the brain can cause hallucinations of gods and spiritual realms and also hyper-religious frenzy and devotion. These very strongly contribute to a person's belief in the supernatural. Even today in the 21st century, people still know so little about these contributors and about how their brains work that they let their imaginations run riot and their ignorance and gullibility take over as they seek explanations.

The irritatingly everpresent god of the gaps illogical escapism rears its ugly head yet again, as the supernatural is offered as the explanation for the experience rather than the actual demonstrable, reasonable and quite mundane and non-supernatural reason – changes in the brain and/or its chemistry. Things would be quite different, and there would be no religion, if more people actually knew how their brains worked.

These physical changes or damage are not the only means by which the brain can experience a change, sometimes both quite devastating and lifelong. The next section explains this.

## Belief Source 3: Physical Changes Or Damage To The Brain From Nonphysical Causes

Beyond the purely physical dangers to the brain, the brain can be changed or damaged merely by what a person sees and experiences, without a physical impact or a chemical input. These sensory inputs, while nonphysical in nature, can have massive effects on the physical brain. This is not nearly as widely known as it should be, is absolutely not intuitive and it is utterly impossible to know this from what one would call "common sense".

Too many people refuse to believe what they actually can see with their own eyes, much less what they can't see, so this point must be made often and loudly.

Post-traumatic stress disorder, or PTSD, is becoming rapidly understood as a set of thoughts and behaviors that results from, according to the US Department of

Veterans Affairs[129], "… experiencing or witnessing a life-threatening event, like combat, a natural disaster, a car accident, or sexual assault." This means that just the simple act of *observing* a stressful or traumatic event can result in permanent changes to the brain and can be quite debilitating to the affected person. Note that the intensity of the event is not on a fixed scale.

What gives one person permanent and debilitating PTSD may not trouble another person at all. Seeing a pile of dead bodies blown to bits may make one person joke about jigsaw puzzles and anatomy classes, while the person next to him may literally freeze up and be unable to function, then suffer lifelong severe PTSD. The problem of PTSD had been observed but not understood or well-researched for quite some time, up until the recent high-level focus on it as part of the treatment of veterans who served in our most recent wars in Iraq and Afghanistan.

I am a retired military veteran and have served in combat zones. I personally know extraordinary and wonderful people who have lost limbs or eyes as a result of being blown up in a combat zone, but who have not suffered marked or debilitating PTSD. I know others who have been blown up on multiple occasions and/or seen massive death and destruction up close but who have no apparent or significant PTSD. I know people who have experienced what most people would call relatively minor stress yet have nevertheless developed PTSD.

Some people have a much lower threshold for dealing with such observations due to the physical construction of their brains. It is not under their control, just as their System 1, hair color, skin color and sexual orientation are not under their control.

Since the massive numbers of people recently affected by Traumatic Brain Injury (TBI) and PTSD due to service in operations in Afghanistan and Iraq made the problem too large and visible to not directly address, there have been formal studies performed by government and civilian entities to learn more about both of these issues.

A 2011 work[130] by Jonathan Sherin and Charles Nemeroff, both medical doctors in the University of Miami's Department of Psychiatry and Behavioral Sciences, is titled "*Post-Traumatic Stress Disorder: The Neurobiological Impact Of Psychological Trauma*" and describes the biology of PTSD.

One of their observations is particularly relevant, and it is that the response to stimulus that results in PTSD is not automatic in everyone, as mentioned above. As already noted, some people's brains are wired such that they are more susceptible to PTSD than others to the same stimuli.

If you lined up 100 people and exposed them to the same stimulus, some will be completely unaffected, others will be mildly upset but not suffer PTSD, some may suffer mild or "unobtrusive" PTSD, and an unlucky few may develop debilitating PTSD. Two very relevant excerpts from the Sherin and Nemeroff work make this biological point:

> *"The signs and symptoms of PTSD, therefore, appear to reflect a persistent, abnormal adaptation of neurobiological*

*systems to the stress of witnessed trauma. The neurobiological systems that regulate stress responses include certain endocrine and neurotransmitter pathways as well as a network of brain regions known to regulate fear behavior at both conscious and unconscious levels."*

*"Efforts to identify neurobiological markers for PTSD originally presumed that abnormalities were acquired "downstream" from an exposure, as a consequence of traumatic experience. It could be, however, that certain abnormalities in the patient with PTSD simply represent pre-existing or "upstream" pathology that is functionally dormant until released by trauma exposure and detected thereafter upon investigation. Along these lines, recent interest has focused on factors that seem to modulate outcome variation in neurobiological systems following trauma exposure including genetic susceptibility factors, female gender, prior trauma, early developmental stage at the time of traumatic exposure, and physical injury (including traumatic brain injury - TBI) at the time of psychological trauma; these parameters likely contribute to vulnerability for, versus resilience against, developing PTSD."*

Some symptoms of PTSD include, according to the US National Institute of Mental Health[131]:

- Flashbacks to the triggering event, and other effects such as perspiring or a rapid pulse.
- Nightmares and/or sleep problems.

- Being scared, fearful, tense, angry, or susceptible to sudden sounds like a car's backfire or a slamming door.
- Not going near or not partaking in reminders of the event, such as not going to certain places or not attending certain events, or avoiding certain people.
- Depression and depressed affect and interest in things.
- Reduced or changed cognition and social skills.

If you have ever been around a service member who served in any of our Nation's combat zones, or even faced threats while serving as a police officer, firefighter, first responder or a similarly stressful profession, you will notice that the symptoms above have been common for some time.

It is often difficult for a person to shoot or stab another person to death, and it is highly unnatural to engage in combat with the weapons and equipment we have invented for that purpose. We can inflict mass death in a matter of seconds, and we can be on the receiving end of deadly attacks. The condition of PTSD is not new. What is new is how much we now know about it.

So the human brain, without being physically struck by an object or force and without being bounced or shaken around inside the skull or exposed to chemicals, can be strongly and permanently affected merely from seeing or experiencing a traumatic event. This is a part of human psychology which every person should know about and which should be taken into consideration as part of medical care, jurisprudence and the criminal justice system, parenting and simply being a human being.

So how do beliefs change after such a traumatic experience? A person who has been in a severe car accident may no longer believe cars or motorized transport is safe. A person who has been raped may no longer believe that men, or men of a certain look or personality, are safe to be around. A child who has been raped or sodomized by a priest may permanently distrust all clergy, even those not wearing Catholic black. A person who has been told and then believes that there is a god who will throw her and her loved ones into an eternal escape-proof hell for any of large a number of acts or even thoughts may be permanently traumatized by this, and may suffer lifelong existential and agonizing guilt, shame and fear due to this belief.

Such beliefs may trigger uncontrollable attempts at compensatory actions such as obsessive-compulsive behavior. Beliefs matter a great deal, and experiences change beliefs.

The fourth and last source of beliefs is the one that we get from the pressures and information we receive as we go through our lives. These don't necessarily rise to the level of producing PTSD, but they do strongly influence beliefs. Social pressures are critical components of a human's life and people start, stop or change their beliefs simply due to social pressure, whether from the need for acceptance or the fear of rejection. Experiences and social inputs and pressures are strong contributors to behavior. Peer pressure and social pressure often trump almost everything else in a person's life, including morality and reason.

## Belief Source 4: How The Brain Responds To Experiences And Sensory Input.

The growth and development of the brain are like the manufacturing processes used to make a computer. If all the right parts are there, if the parts themselves are made of the right elements and parts, if the parts don't have any defects, if the parts are all installed correctly, and if all the manufacturing processes are done properly, then at the end there will be a well-formed computer. The end product will function properly. If some parts are missing, defective or assembled incorrectly, or if any extra parts are added or errors are made during the manufacturing process, the computer will be damaged or unusable. We have already established this.

Once the computer hardware is assembled and it is outfitted with peripherals and basic operating system software, the computer is ready for work, i.e. processing digital data in specific ways through the use of more software which will be added later.

The process of adding software and information on top of the base physical structure and operating system is clean and direct in a computer, but not at all clean or direct in the human brain. The software is actually partially formed and modified as a result of the information the brain receives. A person's brain receives information from the world through the physical senses. This information is not handled in the cold and unthinking manner by which a computer handles information. It is processed based on a wide range of factors which change over time and due to current conditions.

If you try to teach a baby advanced math or physics, the baby will hear your voice but will understand nothing you are saying. If you scream the words at the baby in a loud and angry voice, the baby will still not understand the words but will be quite negatively affected by your tone and volume and will almost certainly begin crying in fear. If you "coo" the exact same words in a loving tone, the baby will still fail to understand the information but the child will respond positively to your pleasant voice and tone.

Whether you screamed at or cooed at the baby, you will have transferred no math or science content but you will have definitely sent it an understandable and very strong emotional message, which it will remember and which it will incorporate into its software. If the baby hears your voice in the future, this will engender a positive feeling if you cooed and a negative feeling if you screamed. Perhaps your deodorant or cologne, your gender, your hair color or your body size will also be associated with cooing pleasantness and calm or screaming hatefulness and fear. The brain is a complex machine and there are numerous ways you will be perceived and how the response patterns will form, in an iterative learning process.

If you as a parent teach your child from birth to his 18$^{th}$ birthday that some group of people, such as Jews, liberals, intellectuals, immigrants or atheists are evil and hateful creatures which are a danger to morality, life, liberty and the pursuit of happiness and deserve to be treated like vermin, you are both adding information to the child's brain and modifying the formation of software and response patterns in that developing brain.

Humans don't imprint[132] in the same way that Konrad Lorenz's goslings did, but humans as social creatures absolutely do strongly model their thinking based on the guidance and examples of their parents and society. There are plenty of contrarians, particularly among teenagers, but by and large the growing child soaks in what his parents and close-in society tell it to do, say and believe, and once inculcated these may be permanent fixtures which can never be removed or changed.

Like a computer virus which may permanently change and damage the processing and outputs of a computer, viral infections of incorrect and/or immoral words spoken to and deeds done to growing children may permanently change and damage the processing and outputs of their brains.

Due to these pressures and the mental viruses transmitted to the young, even an inherently good person with a brain that if left alone would be moral and kind and good can be molded into a callous and immoral person who can engage in very evil words and deeds based on very evil beliefs. The information flow into the growing person's brain is stored as bits of information and also becomes part of the software that develops over time and which forms processes of thinking and beliefs. This has been done so many times in history on scales both large and small that it constitutes one of the most dangerous flaws in human psychology; that a good person can be turned into a bad one, and in far too many ways.

Examples of this include:

- People who were raised to burn heretics and witches during Christian control of Europe.

- People who were raised to believe that torture of infidels or conducting pogroms against Jews was a holy duty.
- People who were told to turn their parents in to the State for "nonconformist" activities, as in Stalinist Russia.
- People who were raised to believe people of other faiths or no faith are lesser beings and/or should be persecuted or treated as less than those of their particular religious in-group.
- People who were raised to believe that people of other colors, languages, appearances, genders and orientations are lesser beings and/or should be persecuted or treated as less human than those of the in-group, or not human at all.

There are many avenues by which people come to hold religious beliefs. Regardless of how wrong or factually insane they may be, beliefs are truly and sincerely held in the mind of the believer. They may be held for the comfort they offer or because they give the believer a mental crutch and external rule or support system he cannot generate internally. Beliefs may be publicly advertised and apparently though not truly held for a variety of reasons. In some cases, being thought to hold certain religious beliefs gives people a societal advantage and/or enables in-group acceptance. In others, faking certain beliefs will stave off persecution and murder at the hands of the most deeply and sincerely faithful believers.

Many people turn to religion as a way of alleviating stress. Humans generally do not like to be highly stressed or to lose their independence, whether that loss results from injury, age or from being subject to the whims of other humans or conditions of nature.

Religion offers the ultimate panacea, that there is a loving god in charge of everything and that no matter how bad things are, everything will be simply *wonderful* one day, whether in this life or in an afterlife. Certain personality types find this psychological escape valve irresistible when their stresses rise above their individual resiliency threshold and they have to turn to artificial external supports, no matter how imaginary those supports are.

Probably the strongest and most telling proof that religion is cultural and societal is the near absence of Muslims in Alabama, the lack of Buddhists in Baton Rouge, and the paucity of Christians in Afghanistan. These conditions demonstrate the extremely local nature of religious belief, which even in a globally interconnected world is still extremely heterogeneous. Surely religious tenets and beliefs, allegedly revealed or otherwise, cannot be true in one place on earth and false in another. Yet billions of people delude themselves in this way.

Familial and social indoctrination and cultural and societal inertia drive the people of one country or locality to toe the party line in their faith and beliefs, particularly when reinforced or even enforced by societal and legal penalties. If the local beliefs are reinforced both in schools and at home, a growing child may not experience any alternative views to the one being preached. If deviating from the beliefs will incur significant cost or pain, the odds of deviation are extremely low and the social inertia will carry the belief for many generations.

Growing up in a particular culture means growing up with certain commonalities. Language is one, religion is another, and even norms such as which system of weights and measures is used are examples of cultural inertia that are highly resistant to change. The United States is almost the only country left on earth that has not adopted the metric system, joining only Burma and Liberia in this dubious condition. Languages and religions persist for hundreds of years. Change is hard, and when change would require changes to infrastructure, voluntary change from within becomes essentially impossible.

As an example, if the US ever adopts the metric system, think of all the housekeeping and infrastructure changes that would need to be made to things as mundane as speed limit signs, highway “mileage” signs, and all the written and electronic references using the old English measurements. It would require a massive amount of money, time and effort, and also Americans just don’t like to be bothered with such things. It has been a much easier and cheaper option to just coast along on the inertia of the original system. Thus we Americans still use miles instead of kilometers.

In the same way, we still use English instead of Chinese or Russian, at least so far, and many of us practice Christianity. Social and cultural inertia are not completely irresistible forces, but getting people to change even obviously bad or immoral ways really is often like trying to move immovable objects.

The bottom line is that in our highly social species, if everyone else is doing it, it is highly likely that you will be doing it too, even if you don’t really want to and even if you don’t actually believe in it. The social costs are just too high to be too far out of the mainstream, and this is true even when you don’t live in a time or place where the ruling religious authorities will actually kill you for being out of that mainstream.

We have now completed the explanation of the four sources of beliefs, which as a reminder are:

1. **The structure and composition of the brain.**
2. **Physical changes or damage to the brain from physical causes.**
3. **Physical changes or damage to the brain from nonphysical causes.**
4. **How the brain responds to experiences and sensory input.**

The above information shows how both nature and nurture are responsible for our beliefs. The structures and processes that lead to a person's beliefs are mostly hard-wired in at birth and are refined via several pathways as the person grows. The human brain has to navigate an immensely challenging and decades-long trek through those twin minefields of nature and nurture to achieve a full and relatively undamaged development. Many of the hazards along the path are out of the person's control, at least until the person is of age and able to be independent and make independent decisions.

## ***So What?***

All the above is quite a lot. But so what? Why go through it all? Because it is extremely important to fully understand how people arrive at then maintain a belief in the supernatural. Only by understanding those processes, like diagnosing and understanding the root cause of a problem, can we first identify then take the appropriate corrective action to bring reason and critical thinking to its proper place as the final arbiter in our dialogue and our actions.

The rest of this book will sum the major points, but the primary point is, yet again, **humanity's invented gods only exist in the imaginations of the believers**, and it is only the widespread lack of knowledge of how the brain works that so many people do not know this.

People often strongly hold their beliefs, no matter how irrational or unjustified, for no reason they can explain. They just "feel that way", which is enough for them to hold their opinion and their religion. This is summed up in that trite expression "if it feels good, do it", but repurposed to "if it feels good, think it". They then reject or twist any and all evidence that disproves their belief or makes them uncomfortable.

Almost no one has any idea how the brain works or why people think what they do. Everyone has deep-seated cognitive biases which drive them to keep thinking the way they do or in ways that make them feel good. People ignore, twist or manipulate facts as needed when maintenance of an opinion requires denial or twisting of reality. This is ubiquitous in religion, because its existential claims are all thoroughly and totally devoid of proof.

It can be and often is impossible to change certain minds even in the complete absence of supporting evidence for a belief, and in a massively perverse logical breakdown, believers often double down and become more fervent believers when they are strongly challenged and shown to be wrong. This is terribly unfortunate, as unjustified beliefs have never resulted in good things for humanity. Part of being a responsible human being is acknowledging that opinions can start a debate but actual facts and actual evidence must always win the debate. This **never** happens in religion.

People don't normally change their minds or beliefs unless there is a significant benefit for changing or a significant penalty for not changing. Even if there is a significant penalty, even all the way to death, many people *still* will not change their minds. And when people do manage to change their beliefs, this is almost always caused by intense external pressures or a major psychological impact, not through a conscious and objective analysis of the relevant facts. Changes through reflection and analysis are the microscopically rare exception. This mental rigidity is extremely common in religion.

The level of investment is critical in the calculus of belief. If nobody cares if you switch from Islam to Catholicism to Satanism to paganism to atheism in the course of a year of "seeking", then you will see people doing this. If a young person has been hiding her homosexuality, atheism or liberalism for years, and knows that due to the Catholic, Baptist, Mormon, Muslim or other beliefs of her family that admitting this will result in being immediately booted from the home, the person will do all she can to keep it hidden.

Most humans can be shaped into good or bad behavior patterns. People born with what would be decent and moral personalities if they are left alone or led to develop that way can unfortunately be molded into people who think and do hateful, violent, vicious and evil things. People with evil tendencies can perhaps not completely eliminate their desires to be evil, but through parental, familial, societal and medical means these desires can be attenuated or judicially dealt with if expressed.

Humans who are born as good, honest, decent and moral people have the genes that make them this way. They tend to marry similar people. These generally good people tend to have children who inherit these genetic traits and personalities. When these kinds of parents raise these children, they and the people they associate with tend to inculcate these positive values in their children. Thus the cycle of moral brains tends to continue, unless the brain is damaged or affected as discussed above.

Humans who are born as bad, dishonest, indecent and immoral people have the genes that make them this way. They also tend to marry similar people, unless they marry good and decent people to take advantage of their spouse and his or her support and resources, as large numbers of them do. These generally bad people tend to have children who inherit these bad genetic traits and personalities. When these kinds of parents raise their children, they and the people they associate with tend to inculcate these negative values in their children. Thus the cycle of negative and potentially antisocial brains tends to continue.

While we cannot precisely determine what percentage of nature or nurture is responsible for a person's behavior, an estimate of 50% for each is a decent starting point. Born psychopaths are not likely to turn into prosocial philanthropist social workers, and vice versa. While it is unfortunate that due to our genetic primate inheritance it is easier to turn a good person into a bad one, there are people born with extremely high levels of morality and goodness who can never be convinced to take immoral action. These people are precious and rare.

The hardware and software in the human brain worked very well for members of our species who lived millions to thousands of years ago. Modern technology and the volume of information available to us today have moved and accumulated at lightspeed compared with the glacially slow pace of our evolution.

Thus our brains are like a primitive 1980s-era computer with a dial-up modem trying to run modern high-speed gaming programs on a super-high-definition display. The old processor, memory and modules just don't have the ability to handle the volume and types of incoming data, but that ancient and creaking computer in our heads has to find some way to handle the inputs. It does so by relying on its primitive basic operating system and a few primitive modules, which frequently give completely wrong answers when used in the modern world.

Our modern brains have pre-Stone Age programming but no modules for an intuitive or "common sense" understanding of the properties of atoms, electricity, gravity, quantum mechanics or so much else in our world, like probability and statistics. We don't have a module that explains intuitively how light behaves as a wave or particle. We don't have a module that intuitively grasps evolution through natural selection. Humanity didn't find out the truth of any of these ideas until our knowledge and science had enabled our partly evolved, non-scientifically inclined, System 1-driven brains to reach the required level of knowledge and understanding.

And even then, breakthroughs like these only came to the brains of a very few extremely intelligent and learned individuals working in those particular fields.

Here's an extremely sobering thought for you. If you took the 100 most innovative, intelligent and impactful scientists throughout history, such as Democritus, Copernicus, Galileo, Darwin, Newton, Einstein, Faraday, Edison, Tesla, Curie, Archimedes and others, and erased them and their contributions from history, our world would be a far different place. We would almost certainly still be reading scripture by candlelight and many people would still be trying to avoid being burned for witchcraft or heresy.

In a similar vein, if you took the 100 most powerful and influential clerics throughout history and removed them and their effects on the world, we might have landed on the moon by the year 1600, cured cancer and most all other afflictions by 1700 and eliminated poverty and human suffering by 1800, not to mention avoiding the uncountable millions of murders that have been carried out by the faithful.

Our brains also don't have a "scientific method" module, and we have struggled mightily as a result. We had to develop the scientific process and the tools to apply it. The method is now in use everywhere in the world and it is working very hard and producing incredible results, but it is being fought tooth and nail by those who want to keep humanity in the first or seventh centuries. That is a huge problem, and is another extremely important "so what".

Religion formed thousands of years ago during a period when that primitive 1980s mental computer was the only processor and assessment tool we had, and the only data we had was what we could sense using our bodies and a few other bits handed down in oral tradition or crude written records. We knew absolutely nothing about the real facts of the world. Yet many people today still cling desperately to and even fight and kill for what ancient ignorant men wrote down, in the face of absolutely overwhelming evidence that almost everything they wrote was wrong. That is yet another very important "so what".

There are other "so what" reasons:

- We have to understand that just as some people find the idea of worshiping some god as crazy an idea as there is, others find it an absolutely irresistible compulsion. Again, we have to understand this and the reasons for it in order to solve this terrible problem.

- Dogma destroys debate, it destroys truth and it never leads to positive outcomes. The dogma of religion is the most powerful justification for evil behavior in our world, and it has been thus ever since the first religion was invented. We have to eliminate dogma and unjustified beliefs. We must ensure our worldviews are proportional to the evidence and that they conform to science. Only then will we be able to live in a truly safe and moral world.

- You really can turn a good person into a bad person. Religion is one of the easiest ways to do this, and for numerous lessons in how to be an extremely bad person all one need do is follow any of the monotheistic holy books. Therefore all unjustified beliefs and ancient superstitions must be blocked and attacked until they cease to exist. The physicist Steven Weinberg said[133] what people have known for millennia. "Religion is an insult to human dignity. With or without it you would have good people doing good things and evil people doing evil things. But for good people to do evil things, that takes religion."

- You really can make a great society if you work at it, but to do so you have to make reason and critical thinking mandatory requirements, and you have to get rid of childish superstition built on ancient ghost stories.

Weinberg had it right, and science and reason get it right. Science and reason are the lights which will illuminate humanity's path to a safer, happier and more moral existence. Religion and superstition blind the mind and darken that path.

Religion and superstition keep Pandora's Box of unjustified dogmas wide open, letting these antihuman pains and Troubles loose on the world to subvert science, suppress reason and spread strife, persecution and immorality wherever they go. Science slams that evil box shut.

Our species will continue to suffer terribly at the hands of religion and superstition until we grow up and once and for all eliminate these poisons from our minds, and then our world. Reason must prevail.

# About The Author

Max Humana is a retired US military servicemember who still supports and defends the Constitution against all enemies both foreign and domestic through writing on the never-safe subjects of religion, politics, morality, humanity, truth and reason.

He is working to advance human knowledge and to rid the world of ignorance, stupidity and evil, and wants you to do the same.

# References


1. Video clip from *The Sixth Sense*. Web resource at https://www.youtube.com/watch?v=QUYKSWQmkrg
2. Pew Research Center publication titled "*Christians remain world's largest religious group, but they are declining in Europe*", by Conrad Hackett and David McClendon, published April 5, 2017. Web resource at http://www.pewresearch.org/fact-tank/2017/04/05/christians-remain-worlds-largest-religious-group-but-they-are-declining-in-europe/
3. Pew Research Center publication titled "*The Global Catholic Population*", dated February 13, 2013. Web resource at http://www.pewforum.org/2013/02/13/the-global-catholic-population/ - how-has-the-population-changed-in-heavily-catholic-countries-in-the-past-decade
4. Time Labs publication titled "*See The Change In Catholic Population Around The World*", dated September 21, 2015. Web resource at http://labs.time.com/story/catholic-population/
5. Sloman, Steven and Fernbach, Philip. *The Knowledge Illusion: Why We Never Think Alone*. New York: Riverhead Books, 2018.
6. *Time Magazine* Interview with Clint Eastwood, article contacts Ty Trippet and Nicole Willner, dated February 20, 2005. Web resource at http://content.time.com/time/press_releases/article/0,8599,1029815,00.html
7. Web resource at https://www.britannica.com/science/DNA
8. Oklahoma State University's Department of Agricultural Economics Food Demand Survey. Web resource at http://www.agecon.okstate.edu/agecon_research.asp
9. Oklahoma State University's Department of Agricultural Economics Food Demand Survey, Volume 2, Issue 9: January 16, 2015. Web resource at http://agecon.okstate.edu/faculty/publications/4975.pdf

10. YouTube video clip from the cartoon series *Futurama*, depicting the refusal of the religious community to accept science, in this case the facts of evolution. Web resource at https://www.youtube.com/watch?v=q-RUHhCzgxI
11. YouTube video clip from the cartoon series *Futurama*, depicting the extreme irritation with the reality-denial of the religious community that is shared by many in the empiricist and scientist communities  Web resource at https://www.youtube.com/watch?v=RaY9k6tuLog
12. Video of Oklahoma Senator James Inofe demonstrating his complete scientific ignorance and dogmatism on the floor of the US Senate.  Web resource at https://www.youtube.com/watch?v=4J4JtMxCJ94
13. Dartmouth Medical School *Review of Clinical and Functional Neuroscience*, Chapter 12, copyright/site editor Rand Swenson, DC, MD, PhD, 2006.  Web resource at https://www.dartmouth.edu/~rswenson/NeuroSci/chapter_12.html
14. Hladky, Stephen B. and Margery A. Barrand. "Mechanisms of fluid movement into, through and out of the brain: evaluation of the evidence" *Fluids and barriers of the CNS* vol. 11,1 26. 2 Dec. 2014, doi:10.1186/2045-8118-11-26.  Web resource at https://www.ncbi.nlm.nih.gov/pmc/articles/PMC4326185/
15. Laterra J, Keep R., Betz LA, et al. "Blood—Cerebrospinal Fluid Barrier". In: Siegel GJ, Agranoff BW, Albers RW, et al., editors. *Basic Neurochemistry*: Molecular, Cellular and Medical Aspects. 6th edition. Philadelphia: Lippincott-Raven; 1999.  Web resource at https://www.ncbi.nlm.nih.gov/books/NBK27998/
16. Stiles, Joan and Terry L. Jernigan. "The basics of brain development" *Neuropsychology review* vol. 20,4 (2010): 327-48.  Web resource at https://www.ncbi.nlm.nih.gov/pmc/articles/PMC2989000/
17. Emory University School of Medicine article titled "*Understanding How the Brain Develops*", by Maureen

Mulligan LaRossa, R.N. and Sheena L. Carter, Ph.D. Web resource at http://www.pediatrics.emory.edu/divisions/neonatology/dpc/brain.html
18. Wehrwein, Hakan S., Orer, HS and Barman, Susan M. "Overview of the Anatomy, Physiology, and Pharmacology of the Autonomic Nervous System". In *Compr Physiol.* 2016; 6(3): 1239–1278. Web resource at https://www.ncbi.nlm.nih.gov/pubmed/27347892
19. Web resource at https://www.merriam-webster.com/dictionary/reflex
20. Stanford Children's Health article titled *Newborn-Reflexes*. Web resource at https://www.stanfordchildrens.org/en/topic/default?id=newborn-reflexes-90-P02630
21. Web resource at https://www.merriam-webster.com/dictionary/instinct
22. Link to National Public Radio's *Hidden Brain* main web page. Web resource at https://www.npr.org/series/423302056/hidden-brain
23. Web resource at https://www.amazon.com/Thinking-Fast-Slow-Daniel-Kahneman/dp/0374533555
24. Web resource at https://www.amazon.com/Righteous-Mind-Divided-Politics-Religion-ebook/dp/B0052FF7YM
25. Kahneman, Daniel. *Thinking, Fast and Slow.* New York: Farrar, Straus and Giroux, 2013, 20.
26. Web resource at https://en.wikipedia.org/wiki/M%C3%BCller-Lyer_illusion
27. Kahneman, Daniel. *Thinking, Fast and Slow*, 52.
28. Ibid., 62.
29. Ibid., 114.
30. Ibid., 158.
31. Ibid., 240.
32. Ibid., 284.
33. National Aeronautics and Space Administration article titled "*Will The Universe Expand Forever?*" by NASA/WMAP Science Team, dated January 24, 2014.

Web resource at
https://map.gsfc.nasa.gov/universe/uni_shape.html
34. Haidt, Jonathan. *The Righteous Mind: Why Good People Are Divided By Politics And Religion.* New York: Vintage Books, 2012, 3.
35. Ibid., 4.
36. Ibid., 22.
37. Ibid., 45.
38. Ibid., 45.
39. Web resource at https://en.wikipedia.org/wiki/Anchoring
40. Web resource at https://en.wikipedia.org/wiki/Anthropocentrism - Cognitive_psychology
41. Web resource at https://www.britannica.com/topic/anthropomorphism
42. Web resource at https://en.wikipedia.org/wiki/Attentional_bias
43. Web resource at https://en.wikipedia.org/wiki/Authority_bias
44. Web resource at https://en.wikipedia.org/wiki/Availability_cascade
45. Web resource at https://www.goodreads.com/quotes/7676946-history-is-strewn-thick-with-evidence-that-a-truth-is
46. *Scientific American* article titled "*How to Convince Someone When Facts Fail*", by Michael Shermer, date January 1, 2017. Web resource at https://www.scientificamerican.com/article/how-to-convince-someone-when-facts-fail/
47. Web resource at https://en.wikipedia.org/wiki/Bandwagon_effect
48. Web resource at https://en.wikipedia.org/wiki/Belief_bias
49. Web resource at https://en.wikipedia.org/wiki/Bias_blind_spot
50. Web resource at https://en.wikipedia.org/wiki/Choice-supportive_bias
51. Web resource at https://en.wikipedia.org/wiki/Clustering_illusion

52. Web resource at https://en.wikipedia.org/wiki/Confirmation_bias
53. Web resource at https://en.wikipedia.org/wiki/Curse_of_knowledge
54. Web resource at https://en.wikipedia.org/wiki/Dunning%E2%80%93Kruger_effect
55. Web resource at https://en.wikipedia.org/wiki/Egocentric_bias
56. Web resource at https://en.wikipedia.org/wiki/False_consensus_effect
57. Web resource at https://www.britannica.com/science/Barnum-Effect
58. Web resource at https://en.wikipedia.org/wiki/Framing_effect_(psychology)
59. The University of Texas at Austin McCombs School of Business article titled "*Fundamental Attribution Error*", with video, copyright 2018. Web resource at https://ethicsunwrapped.utexas.edu/glossary/fundamental-attribution-error
60. Web resource at https://www.britannica.com/topic/hindsight-bias
61. Gilbert, E.T., Brown, R.P., Pinel, E.C. and Wilson, T.D. "The illusion of external agency". *J Pers Soc Psychol.* 2000 Nov; 79(5): 690–700. Web resource at https://www.ncbi.nlm.nih.gov/pubmed/11079235
62. Web resource at https://en.wikipedia.org/wiki/Illusory_truth_effect
63. Web resource at https://en.wikipedia.org/wiki/Illusory_correlation
64. Article in *Psychology Today* titled "*In-groups, out-groups, and the psychology of crowds*", by Susan Krauss Whitbourne, Ph.D., dated December 7, 2010. Web resource at https://www.psychologytoday.com/us/blog/fulfillment-any-age/201012/in-groups-out-groups-and-the-psychology-crowds
65. Santa Clara University's Markkula Center for Applied Ethics article titled "*The Just World Theory*", by Claire

Andre and Manuel Velasquez, posted November 13, 2015; originally published in *Issues in Ethics* - V. 3, N. 2 Spring 1990. Web resource at https://www.scu.edu/ethics/ethics-resources/ethical-decision-making/the-just-world-theory/

66. Web resource at https://en.wikipedia.org/wiki/Law_of_the_instrument
67. Article in *Psychology Today* titled "*To know me is to like me I: Mere exposure*", by Art Markman, Pd.D., dated November 20, 2008. Web resource at https://www.psychologytoday.com/us/blog/ulterior-motives/200811/know-me-is-me-i-mere-exposure
68. Web resource at https://en.wikipedia.org/wiki/Na%C3%AFve_realism_(psychology)
69. Web resource at https://en.wikipedia.org/wiki/Ostrich_effect
70. Web resource at https://en.wikipedia.org/wiki/Overconfidence_effect
71. Web resource at https://www.merriam-webster.com/dictionary/pareidolia
72. Article in *Scientific American* titled "*The Believing Brain: Why Science Is the Only Way Out of Belief-Dependent Realism*", by Michael Shermer, dated July 1, 2011. Web resource at https://www.scientificamerican.com/article/the-believing-brain/
73. Web resource at https://www.merriam-webster.com/dictionary/post hoc, ergo propter hoc
74. Web resource at https://en.wikipedia.org/wiki/Choice-supportive_bias
75. Web resource at https://en.wikipedia.org/wiki/Selective_perception
76. Article in *Psychology Today* titled "*Selective Perception and Attention at the World Cup*", by Alan Castel, Ph.D., dated June 26, 2014. Web resource at https://www.psychologytoday.com/us/blog/metacognition-and-the-mind/201406/selective-perception-and-attention-the-world-cup

77. Article in *Psychology Today* titled "*The Self-Serving Bias - Definition, Research, and Antidotes*", by Alice Boyes, Ph.D., dated January 9, 2013. Web resource at https://www.psychologytoday.com/us/blog/in-practice/201301/the-self-serving-bias-definition-research-and-antidotes
78. Web resource at https://en.wikipedia.org/wiki/Semmelweis_reflex
79. Web resource at https://en.wikipedia.org/wiki/Shared_information_bias
80. Web resource at https://en.wikipedia.org/wiki/Status_quo_bias
81. Web resource at https://en.wikipedia.org/wiki/Subjective_validation
82. Web resource at https://dictionary.cambridge.org/us/dictionary/english/sunk-cost-fallacy
83. Article in *Psychology Today* titled "*Letting Go of Sunk Costs*", by Robert L. Leahy, Ph. D., dated September 24, 2014. Web resource at https://www.psychologytoday.com/us/blog/anxiety-files/201409/letting-go-sunk-costs
84. Web resource at https://en.wikipedia.org/wiki/Survivorship_bias
85. Article at PLOS One titled "*Shared Brain Lateralization Patterns in Language and Acheulean Stone Tool Production: A Functional Transcranial Doppler Ultrasound Study*", by Natalie Thais Uomini and Georg Friedrich Meyer, dated August 30, 2013. Web resource at https://journals.plos.org/plosone/article?id=10.1371/journal.pone.0072693
86. Smithsonian Institution National Museum of Natural History article titled "*What does it mean to be human*?", last updated August 24, 2018. Web resource at http://humanorigins.si.edu/evidence/human-fossils/species/homo-sapiens
87. National Center for Biotechnology Information. *PubChem Compound Database*; CID=681, https://pubchem.ncbi.nlm.nih.gov/compound/681

(accessed Nov. 23, 2018). Web resource at https://pubchem.ncbi.nlm.nih.gov/compound/dopamine - section=Top
88. Sathyanarayana Rao, T S et al. "The biochemistry of belief" *Indian journal of psychiatry* vol. 51,4 (2009): 239-41. Web resource at https://www.ncbi.nlm.nih.gov/pmc/articles/PMC2802367/
89. Schmack, Katharina et al. "Linking unfounded beliefs to genetic dopamine availability" *Frontiers in human neuroscience* vol. 9 521. 30 Sep. 2015, doi:10.3389/fnhum.2015.00521. Web resource at https://www.ncbi.nlm.nih.gov/pmc/articles/PMC4588007/
90. Krummenacher P, Mohr C, Haker H, Brugger P., "Dopamine, paranormal belief, and the detection of meaningful stimuli". *J Cogn Neurosci.* 2010 Aug;22(8):1670-81. doi: 10.1162/jocn.2009.21313. Web resource at https://www.ncbi.nlm.nih.gov/pubmed/19642883/
91. Article in *Psychology Today* titled "*Your Sixth Sense*", by Matthew Hutson, published July 3, 2012. Web resource at https://www.psychologytoday.com/us/articles/201207/your-sixth-sense
92. Web resource at https://en.wikipedia.org/wiki/Days_of_Prayer_for_Rain_in_the_State_of_Texas
93. Article in *Scientific American* titled "*Fossil Reveals What Last Common Ancestor of Humans and Apes Looked Like*", by Charles Q. Choi, LiveScience, dated August 10, 2017. Web resource at https://www.scientificamerican.com/article/fossil-reveals-what-last-common-ancestor-of-humans-and-apes-looked-liked/
94. Web resource at https://en.wikipedia.org/wiki/Moore%27s_law
95. Article in *Popular Science* titled "*The incredible evolution of supercomputers' powers, from 1946 to today*", by Sara Chodosh, dated April 22, 2017. Web

resource at https://www.popsci.com/supercomputers-then-and-now
96. Morgane PJ1, Austin-LaFrance R, Bronzino J, Tonkiss J, Díaz-Cintra S, Cintra L, Kemper T, Galler JR., "Prenatal malnutrition and development of the brain". *Neurosci Biobehav Rev.* 1993 Spring;17(1):91-128. Web resource at https://www.ncbi.nlm.nih.gov/pubmed/8455820
97. From the Emory University School of Medicine, Department of Psychiatry and Behavioral Sciences, article by the Maternal Substance Abuse and Child Development Project titled "*The effects of maternal stress and anxiety during pregnancy*", point of contact is Katrina C. Johnson, Ph.D. Web resource at http://www.psychiatry.emory.edu/PROGRAMS/GADrug/Feature Articles/Mothers/The effects of maternal stress and anxiety during pregnancy (mot07).pdf
98. Kinsella, Michael T and Catherine Monk. "Impact of maternal stress, depression and anxiety on fetal neurobehavioral development" *Clinical obstetrics and gynecology* vol. 52,3 (2009): 425-40. Web resource at https://www.ncbi.nlm.nih.gov/pmc/articles/PMC3710585/
99. Bublitz, Margaret H and Laura R Stroud. "Maternal smoking during pregnancy and offspring brain structure and function: review and agenda for future research" *Nicotine & tobacco research: official journal of the Society for Research on Nicotine and Tobacco* vol. 14,4 (2011): 388-97. Web resource at https://www.ncbi.nlm.nih.gov/pmc/articles/PMC3313781/
100. Posecion, N et al. "Detection of Exposure to Environmental Pesticides During Pregnancy by the Analysis of Maternal Hair Using GC-MS" *Chromatographia* vol. 64,11-12 (2006): 681-687. Web resource at https://www.ncbi.nlm.nih.gov/pmc/articles/PMC1934610/
101. Mayo Clinic article titled "*Fetal alcohol syndrome*", by the Mayo Clinic Staff, dated January 10, 2018. Web

resource at https://www.mayoclinic.org/diseases-conditions/fetal-alcohol-syndrome/symptoms-causes/syc-20352901

102. Article by the Division of Birth Defects and Developmental Disabilities, NCBDDD, Centers for Disease Control and Prevention titled "*Facts about Anencephaly*", page last updated on August 2, 2017. Web resource at https://www.cdc.gov/ncbddd/birthdefects/anencephaly.html
103. Article by the National Center for Emerging and Zoonotic Infectious Diseases (NCEZID), Division of Vector-Borne Diseases (DVBD), Centers for Disease Control and Prevention, titled "*Pregnancy*", page last updated November 6, 2017. Web resource at https://www.cdc.gov/zika/pregnancy/index.html
104. Article by the National Center for Emerging and Zoonotic Infectious Diseases (NCEZID), Division of Vector-Borne Diseases (DVBD), Centers for Disease Control and Prevention, titled "*Microcephaly & Other Birth Defects*", page last updated March 1, 2018. Web resource at https://www.cdc.gov/zika/healtheffects/birth_defects.html
105. Article by the University of Akron, from the Phineas Gage information page, titled "*The Phineas Gage story*", created by Malcolm Macmillan of the School of Psychology, Deakin University; School of Psychological Sciences, University of Melbourne, Victoria. Australia and now maintained by The Center for the History of Psychology at The University of Akron, Ohio. Web resource at https://www.uakron.edu/gage/story.dot
106. Article from *Smithsonian Magazine* titled "*Phineas Gage: Neuroscience's Most Famous Patient*", by Steve Twomey, January 2010. Web resource at https://www.smithsonianmag.com/history/phineas-gage-neurosciences-most-famous-patient-11390067/
107. Article from National Public Radio titled "*Why Brain Scientists Are Still Obsessed With The Curious Case Of*

*Phineas Gage*", by Jon Hamilton, dated May 21, 2017. Web resource at https://www.npr.org/sections/health-shots/2017/05/21/528966102/why-brain-scientists-are-still-obsessed-with-the-curious-case-of-phineas-gage

108. Web resource at https://www.merriam-webster.com/dictionary/hallucination
109. McGrath JJ, Saha S, Al-Hamzawi A, et al. Psychotic Experiences in the General Population A Cross-National Analysis Based on 31 261 Respondents From 18 Countries. *JAMA Psychiatry*. 2015;72(7):697–705. doi:10.1001/jamapsychiatry.2015.0575. Web resource at https://jamanetwork.com/journals/jamapsychiatry/fullarticle/2298236
110. Teeple, Ryan C et al. "Visual hallucinations: differential diagnosis and treatment" *Primary care companion to the Journal of clinical psychiatry* vol. 11,1 (2009): 26-32. Web resource at https://www.ncbi.nlm.nih.gov/pmc/articles/PMC2660156/
111. Ibid.
112. Mayo Clinic article titled "*Lewy body dementia*", by the Mayo Clinic Staff. Web resource at https://www.mayoclinic.org/diseases-conditions/lewy-body-dementia/symptoms-causes/syc-20352025
113. Article from the United Kingdom's National Health Service, titled "*Charles Bonnet syndrome*", page last reviewed February 15, 2016. Web resource at https://www.nhs.uk/conditions/charles-bonnet-syndrome/
114. ScienceDirect resource "Anton-Babinski syndrome". Web resource at https://www.sciencedirect.com/topics/medicine-and-dentistry/antonbabinski-syndrome
115. Web resource at https://en.wikipedia.org/wiki/Peduncular_hallucinosis
116. Sahoo, Swapnajeet et al. "Acute Hyperglycemia Associated with Psychotic Symptoms in a Patient with Type 1 Diabetes Mellitus: A Case Report" *Innovations in clinical neuroscience* vol. 13,11-12 25-27. 1 Dec.

2016. Web resource at https://www.ncbi.nlm.nih.gov/pmc/articles/PMC5300708/

117. Article at Huffpost.com titled "*Not Quite Enough: The Consequences of Sleep Deprivation*", by Brandon R. Peters, M.D., posted July 11, 2013 and updated September 10, 2013. Web resource at https://www.huffpost.com/entry/sleep-deprivation_b_3536674
118. Article in *Scientific American* titled "*The Sensed-Presence Effect*", by Michael Shermer, dated April 1, 2010. Web resource at https://www.scientificamerican.com/article/the-sensed-presence-effect/
119. Article from National Public Radio titled "Guardian Angels Or The 'Third Man Factor'?", from the program *All Things Considered*, dated September 13, 2009. Web resource at https://www.npr.org/templates/story/story.php?storyId=112746464.
120. Web resource at https://www.amazon.com/Third-Man-Factor-Surviving-Impossible/dp/1602861293
121. Article at abc.net.au titled "*Finding God in a seizure: the link between temporal lobe epilepsy and mysticism*", by Jess Hill, dated December 10, 2014. Web resource at https://www.abc.net.au/radionational/programs/archived/encounter/the-link-between-temporal-lobe-epilepsy-and-mysticism/5956982
122. Landtblom AM, Lindehammar H, Karlsson H, Craig AD, "Insular cortex activation in a patient with "sensed presence"/ecstatic seizures". *Epilepsy Behav.* 2011 Apr;20(4):714-8. doi: 10.1016/j.yebeh.2011.01.031. Epub 2011 Mar 25. Web resource at https://www.ncbi.nlm.nih.gov/pubmed/21440512
123. Article at LiveScience.com, titled "*Near-Death Experiences Explained by Science*", by Charles Q. Choi, September 12, 2011. Web resource at https://www.livescience.com/16019-death-experiences-explained.html

124. Article by the National Institute on Drug Abuse; National Institutes of Health; U.S. Department of Health and Human Services titled "*What are hallucinogens?*", revised January 2016. Web resource at https://www.drugabuse.gov/publications/drugfacts/hallucinogens
125. Web resource at https://en.wikipedia.org/wiki/Dopamine_agonist
126. Article by the National Institute on Drug Abuse; National Institutes of Health; U.S. Department of Health and Human Services titled "*What are marijuana effects?*" Web resource at https://www.drugabuse.gov/publications/research-reports/marijuana/what-are-marijuana-effects
127. Web resource at https://en.wikipedia.org/wiki/MDMA
128. YouTube video clip of Beavis's absolutely hilarious hallucinogenic trip from the movie *Beavis and Butt-Head Do America*. Web resource at https://www.youtube.com/watch?v=xo61f8aXrbg
129. Reference from the U.S. Department of Veterans Affairs and the National Center for PTSD. Web resource at https://www.ptsd.va.gov/
130. Sherin, Jonathan E and Charles B Nemeroff. "Post-traumatic stress disorder: the neurobiological impact of psychological trauma" *Dialogues in clinical neuroscience* vol. 13,3 (2011): 263-78. Web resource at https://www.ncbi.nlm.nih.gov/pmc/articles/PMC3182008/
131. Article by the National Institute of Mental Health; National Institutes of Health; U.S. Department of Health and Human Services titled "*Post-Traumatic Stress Disorder*". Web resource at https://www.nimh.nih.gov/health/topics/post-traumatic-stress-disorder-ptsd/index.shtml
132. Web resource at https://www.britannica.com/topic/imprinting-learning-behaviour
133. Quote from Steven Weinberg. Web resource at https://en.wikiquote.org/wiki/Steven_Weinberg

Made in United States
North Haven, CT
04 November 2026

21765612R00102